Drone Piloting

by Curt Simmons

for dummies®
A Wiley Brand

Drone Piloting For Dummies®

Published by: **John Wiley & Sons, Inc.**, 111 River Street, Hoboken, NJ 07030-5774, www.wiley.com

For general information on our other products and services, please contact our Customer Care Department within the U.S. at 877-762-2974, outside the U.S. at 317-572-3993, or fax 317-572-4002. For technical support, please visit https://hub.wiley.com/community/support/dummies.

Wiley publishes in a variety of print and electronic formats and by print-on-demand. Some material included with standard print versions of this book may not be included in e-books or in print-on-demand. If this book refers to media that is not included in the version you purchased, you may download this material at http://booksupport.wiley.com. For more information about Wiley products, visit www.wiley.com.

Library of Congress Control Number: 2024946990

ISBN 978-1-394-28211-1 (pbk); ISBN 978-1-394-28213-5 (ebk); ISBN 978-1-394-28215-9 (ebk)

SKY10090438_110824

Contents at a Glance

Introduction 1

Part 1: Becoming a Drone Pilot 5
CHAPTER 1: Finding Your Place in the Sky 7
CHAPTER 2: Flying for the First Time 27
CHAPTER 3: Maneuvering and Flying Your Drone 45
CHAPTER 4: Taking Care of Your Drone and Flying Safely 65

Part 2: Following FAA Regulations 77
CHAPTER 5: Exploring the TRUST Certificate and Part 107 Certification 79
CHAPTER 6: Following Flight Regulations 99
CHAPTER 7: Following More Flight Regulations 111
CHAPTER 8: Flying at Night 123

Part 3: Getting to Know the National Airspace System (NAS) and Sectional Charts 141
CHAPTER 9: Checking Out the National Airspace System 143
CHAPTER 10: Getting to Know Confusing Acronyms and Airport Radio Communications 163
CHAPTER 11: Discovering How to Read Sectional Charts 177
CHAPTER 12: Practically Interpreting Sectional Charts 191
CHAPTER 13: Flying Drones around Airports 207

Part 4: Flight Operations and Safety 219
CHAPTER 14: Piloting in Various Weather Conditions 221
CHAPTER 15: Flight Emergency Procedures 233
CHAPTER 16: Becoming a Safe, Responsible Drone Pilot 243

Part 5: Getting to Work: Drone Piloting in the Marketplace 257
CHAPTER 17: Taking Great Drone Photos 259
CHAPTER 18: Shooting Expert Drone Video 277
CHAPTER 19: Getting Drone Piloting Jobs 293

Part 6: The Part of Tens 313

CHAPTER 20: Ten Part 107 Test-Day Tips 315

CHAPTER 21: Ten Drone Piloting Myths 321

CHAPTER 22: Ten Common Drone Piloting Problems You Can Avoid 327

Index 333

Table of Contents

INTRODUCTION 1

About This Book. 1

Foolish Assumptions. 2

Icons Used in This Book 3

Beyond the Book. 3

Where to Go from Here 3

PART 1: BECOMING A DRONE PILOT 5

CHAPTER 1: **Finding Your Place in the Sky**. 7

Exploring Drone Piloting. 8

Piloting as a hobby 8

Piloting as a profession. 8

Should you become a remote pilot?. 10

Looking at How Drones Work 12

How drones fly. 12

Using the remote controller (RC). 15

Understanding camera gimbals 17

Shopping for a Drone 18

Checking out the 250-gram weight maximum 18

Exploring drone categories 19

Shopping tips to keep in mind. 24

Considering Warranties and Insurance 25

Basic and extended warranties. 25

Third-party plans 25

Replacement plans 25

Additional insurance. 26

CHAPTER 2: **Flying for the First Time** 27

Registering Your Drone and Checking for Updates 28

Registering your drone with the FAA 28

Checking for updates 30

Getting Familiar with Your Remote Controller 32

Understanding How GPS Works with Your Drone 34

Downloading a B4UFLY App 35

Using Your Drone's Crash Safety Features 36

Understanding Geofencing 38

Choosing a First Flight Location. 38

Following Pre-Flight Checks and Procedures .39

Taking Off and Landing for the First Time 40
What to Do When You Don't Know What to Do 42
Following Post-Flight Checks 42

CHAPTER 3: **Maneuvering and Flying Your Drone** 45
Exploring Foundational Flight Principles 46
    Safety first 46
    Consider the weather 47
    Always perform pre-flight checks 47
Practicing Basic Drone Flight Maneuvers 47
    Straight up, straight down 48
    Straight forward, straight back 49
    Circle 50
    Rectangle 50
    Triangle 51
    Plus sign 52
    Stairs 53
Pushing Forward with Complex Maneuvers 54
    Spiral 54
    Orbit 55
    Fly with gimbal rotation 56
Flying with Built-in Maneuvers 58
Shooting with Active Tracking 59
Using DJI MasterShots 61
Growing Your Skills and Confidence 63

CHAPTER 4: **Taking Care of Your Drone and Flying Safely** 65
Keeping Current with Updates 66
Inspecting and Changing Propellers 66
    Inspecting propellers 67
    Changing propellers 68
Cleaning Your Drone and Camera 68
    Cleaning the drone body and propellers 68
    Cleaning the drone camera and gimbal 69
Calibrating the Drone 70
Taking Care of LiPo Batteries 71
Troubleshooting Common Problems 73
    Environmental effects 73
    Camera issues 74
    Connectivity glitches 74
    Getting help 74
Exploring Foundational Safety Practices 74
    Maintain visual line of sight 75
    Fly in authorized airspace 75
    Do not fly in a dangerous manner 75

Stay away from emergency scenes and personnel 75
Do not fly under the influence of drugs and alcohol 76
Don't talk to people when you fly . . : 76

PART 2: FOLLOWING FAA REGULATIONS. 77

CHAPTER 5: **Exploring the TRUST Certificate and Part 107 Certification** 79
Clearing the Air about TRUST Certificates and the Part 107 Exam. . . .80
Drone weight 80
Recreational flyers and TRUST certificates 81
Certified remote pilots 82
Determining Whether You Need a Part 107 Certification 83
Beach photos 83
Family reunion 83
Small business photos 84
Nonprofit photos. 85
Friends building a house 85
Insta-famous mountains 86
Checking out the Part 107 Exam. 86
Understanding airspace. 87
Checking out flight regulations 88
Dealing with weather 89
Considering loading and performance .90
Understanding airport operations 91
Understanding Crew Resource Management. 92
Studying for the Part 107 Exam. 94
Registering for the Part 107 Exam 96

CHAPTER 6: **Following Flight Regulations**. 99
Understanding Maximum Height 100
Understanding MSL and AGL. 100
I told you there are two exceptions 102
Exploring Speed and Weather Visibility 103
Keeping a Visual Line of Sight (VLOS) 104
Flying During Twilight 105
Complying with Remote Identification. 106
Standard Remote ID 106
Remote ID broadcast module 107
FRIA exception 108

CHAPTER 7: **Following More Flight Regulations** 111
Flying Over Property and People 112
Flying over property 112
Flying over people 113
Flying over large groups of people 116

Flying from a Moving Vehicle..............................117
Reporting an Accident to the FAA..........................117
Logging Your Flights and Maintenance 118
Obtaining Airspace Authorization or Waivers...............119
    Getting airspace authorization 119
    Getting LAANC authorization...........................120
Understanding the Effects of Alcohol and Drugs 121

CHAPTER 8: **Flying at Night**..............................123
Exploring FAA Regulations for Night Flights 124
    Flying over people and moving vehicles at night 124
    Airspace and LAANC waivers............................124
    Anti-collision lighting...............................125
Understanding Drone Limitations with Night Flights........126
    Camera limitations 126
    Sensor limitations 127
    Battery limitations..................................128
Considering Eyesight Limitations at Night.................128
    Understanding the basics of your eyesight.............129
    Exploring rods and cones.............................130
    Adapting to dark conditions..........................132
Understanding Night-Flight Visual Problems and Illusions........133
    Phantom motion 133
    Fascination 133
    Autokinesis..133
    Flicker vertigo......................................134
    Size-distance illusion...............................134
    Reversible perception illusion 134
Developing Night-Flight Best Practices....................136
    Check out the area in the daytime 137
    Don't ignore pre- and post-flight checks 137
    Don't forget LAANC approval if needed................137
    Use off-center viewing and scanning 138
    Be physically and mentally ready 139

PART 3: GETTING TO KNOW THE NATIONAL
AIRSPACE SYSTEM (NAS) AND SECTIONAL CHARTS 141

CHAPTER 9: **Checking Out the National Airspace System** 143
Class A Airspace..144
Class B Airspace..145
Class C Airspace..149
Class D Airspace..151

Class E Airspace 153
    Starting from the ground up 154
    Starting from 700 feet AGL. 155
    Starting from 1,200 feet AGL 157
Class G Airspace. 157
Types of Special-Use Airspace 157
    Warning. 158
    Restricted 158
    Prohibited. 159
    Alert 159
    Military Operations Area (MOA) 160

CHAPTER 10: **Getting to Know Confusing Acronyms and
Airport Radio Communications** 163
Understanding Sectional Charts 164
Exploring Common Acronyms and Notations. 165
    NOTAM 165
    TFR. 166
    MTR 167
    Victor Airway 170
    VFR Checkpoint 172
Deciphering Airport Radio Communications. 172
    UNICOM and CTAF 173
    CT and ATIS. 174

CHAPTER 11: **Discovering How to Read Sectional Charts** 177
Understanding Latitude and Longitude. 178
    Reading locations on a sectional chart. 179
    Understanding GPS digital coordinates .182
Touching on Isogonic Lines 183
Understanding Elevation 183
    Natural elevation maximums 184
    Maximum Elevation Figures (MEF) 184
    Towers and other obstructions 185
Pop Quiz! Bringing Everything Together 186

CHAPTER 12: **Practically Interpreting Sectional Charts**. 191
Zoning In on Questions about Airspace and Airports 192
    Testing supplement: Figure 20 192
    Testing supplement: Figure 21 194
    Testing supplement: Figure 25 195
Focusing on Elevations and Locations 197
    Testing supplement: Figure 26 197

Dodging Towers and Obstructions 199
   Testing supplement: Figure 24 199
   Testing supplement: Figure 25 201
Staying Alert to Special-Use Areas 202
   Testing supplement: Figure 75 202
   Testing supplement: Figure 69 204

CHAPTER 13: **Flying Drones around Airports** 207
Reading Chart Supplements 208
Understanding Towered and Nontowered Airports 210
Exploring Airport Traffic 211
   The rundown on how runways work .211
   Understanding runway markings 214
   Knowing how planes land 214
Understanding Right of Way 218

PART 4: FLIGHT OPERATIONS AND SAFETY 219

CHAPTER 14: **Piloting in Various Weather Conditions** 221
Exploring Air Masses. 222
   Connecting cloud types with weather systems. 222
   Getting current on stable and unstable air 224
Piloting through Convection Currents 225
Dealing with the Wind. 226
   Flying with obstructions in mind. 226
   Maneuvering through wind shear. 226
   Staying up in downdrafts 227
Watching Out for Humidity and Fog. 227
   Working harder in low and high humidity 227
   Figuring out fog and visibility. 228
Understanding METAR and TAF Reports 229
   METAR. 229
   TAF. 230
Tips for Flying in Bad Weather. 232

CHAPTER 15: **Flight Emergency Procedures** 233
Looking at Common Drone Emergency Causes 234
   Lost link. 234
   Weather 234
   Batteries 235
   Birds 235
   Overdependance on obstacle avoidance. 236
   Carelessness. 236
Practicing Foundational Emergency Responses 237
Understanding Lost-Link Procedures. 238
   Knowing why lost links occur. 238
   Dealing with a lost link 239

CHAPTER 16: **Becoming a Safe, Responsible Drone Pilot** 243

    Understanding Loading 244
        Force equilibrium 244
        Figuring load factor 245
        Taking center of gravity into account 247
        Understanding the critical angle of attack for
        fixed-wing drones 248
    Making Sound Flight Judgment Calls 249
    Considering Other Models 250
        DECIDE model 250
        IMSAFE model 251
        PAVE model 252
    Getting Rid of Incorrect Thinking 252
        Anti-authority 253
        Impulsivity 253
        Invulnerability 254
        Machoism 254
        Resignation 255

**PART 5: GETTING TO WORK: DRONE PILOTING
IN THE MARKETPLACE** 257

CHAPTER 17: **Taking Great Drone Photos** 259

    Considering the Basics of Good Drone Photography 260
        Understanding shooting modes 261
        Focusing on Exposure 262
    Considering File Types and Resolution 266
        Drone file types 266
        Photo resolution 267
    Shooting Good Angles and Using ND Filters 268
        Applying composition 268
        Shooting with good angles 269
        Working with ND filters 273
    Editing Drone Photos 274
        Consider cropping 274
        Improving exposure 275
        Improving color 275

CHAPTER 18: **Shooting Expert Drone Video** 277

    Filming Great Drone Video 278
        Make a plan 278
        Consider the audience 278
        Modify on the fly 279
        Deliver the details 279
        Shoot three times what you need 279

Understanding File Formats, Resolution, and FPS 280
    File formats. 280
    Resolution. 281
    FPS 281
Shooting with the Best Drone Movements .282
    Ascend and descend. 282
    Orbit 283
    Semicircle 284
    Bird's-eye rotation. 285
    Fly in/pull back. 285
    Sweep 287
    Track 287
Getting Started with Video Editing 288
    Importing video clips. 289
    Editing each clip on the timeline. 289
    Arrange clips and apply transitions 290
    Edit for color and exposure 291
    Add music, voiceovers, and other effects 292
    Finalize, export . . . and then review! 292

CHAPTER 19: **Getting Drone Piloting Jobs** 293
Exploring Drone Piloting in Business 294
    Real estate and mapping 294
    Agriculture 296
    Law enforcement. 297
    Search and rescue. 298
    Construction and inspections 299
    Shipping and logistics 300
    Film and photography 301
    Security 301
Finding Your Drone Piloting Career 302
    Looking at your current job 302
    Exploring your area of interest 303
    Developing skills as you work 304
Setting Up a Drone Piloting Freelance Business. 304
    Defining your business. 305
    Setting up your business 305
    Making first contact. 306
    Establishing an internet presence. 307
    Advertising your business 308
    Getting paid 309
    Juggling income and expenses 311
    Considering drone insurance 311
    Paying your dreaded taxes. 312

PART 6: THE PART OF TENS.................................313

CHAPTER 20: Ten Part 107 Test-Day Tips.........315
Eat, Sleep, Repeat 315
Buy a Magnifying Glass and Cheap Calculator 316
Make Notes Before You Start 316
Slow Down 317
Increase Your Odds...............................317
Read at "Face Value" 318
Mark Tricky Questions 318
Quickly Guess at Questions You Don't Know 319
Dress for Success................................319
Know the Test Details 320
Remember That the World Won't End 320

CHAPTER 21: Ten Drone Piloting Myths.................321
You Don't Need Certification or Registration If Your
Drone Is Small 321
Crewed Aircraft Do Not Fly under 400 Feet.................322
You Don't Have to Worry About Airspace as Long as You
Fly under 400 Feet..............................323
You Don't Need Line of Sight If Your Drone Has a Good Camera 323
You Don't Have to Report an Accident to the FAA 324
You Don't Have to Worry About MTRs 324
Elevation Doesn't Impact Drone Battery Life.................325
After You Learn How to Fly, You Don't Need RTH................325
The Built-In Drone Lights Are All You Need to Fly at Night 326
Drone Piloting Jobs Don't Require State Sales Tax...............326

**CHAPTER 22: Ten Common Drone Piloting Problems
You Can Avoid**...................................327
Software Glitches and Errors.........................327
Shaky Video and Blurry Images........................328
Flying in the Wrong Airspace 328
Connectivity Problems with the Drone and RC................329
Overestimating Battery Life..........................329
Losing VLOS 330
Flights Seem Tilted or Off Balance 330
Poor GPS Connectivity 330
Overdependance on Obstacle Avoidance 331
You...331

INDEX...333

Introduction

First it's the sound — that distinctive loud, buzzy droning noise — that sparks your excitement. Then you suddenly realize that even though you're standing on the ground, you hold in your hands the power to see the world around you — from the sky! As you direct the aircraft to rise above the trees or buildings around you and move in any direction you want, it's like going on an adventure. And it's like that every single time you fly.

If this description of flying a drone sounds tempting, drone piloting may be calling your name!

Flying a drone is both challenging and thrilling, but it also increasingly offers career possibilities. Maybe you've picked up this book because you're just getting started with drone piloting as a hobby. Or maybe you have an established hobby, such as photography, but are thinking about doing more. You may even be a semi-professional drone pilot who wants to take things to the next level with a career or small business. Drones fulfill many important purposes these days, with more uses for them evolving all the time. And besides, drone piloting is just fun!

Becoming a drone pilot isn't easy. It takes time to learn to fly a drone safely, effectively, and within Federal Aviation Administration (FAA) rules when they apply, and gathering all the information needed to do that isn't the fun part of drone piloting. The internet is full of information, but sometimes (and, on social media, often) you find information that's contradictory or simply wrong.

Drone Piloting For Dummies is here to help you fly above a sea of information to find out what you need to know.

About This Book

This book answers pretty much all your basic questions about drones, including how they work, how to figure out which drone is best for your needs, which rules of flying apply to you, and where and under what conditions you can legally fly. This book also helps to prepare you to move beyond hobbyist to professional drone pilot if that's your goal.

I worked hard to make this book easy to understand. I cover beginning topics like flying a drone for the first time, but you also delve into airspace and how to understand FAA sectional charts. Grasping these concepts involves a lot of technical details, but I present them clearly and concisely, with quite a few illustrations along the way to enhance understanding. If you want to do anything commercially with a drone, you need to pass the Part 107 exam of the FAA. Having the Part 107 certification enables you to fly legally when you engage in for-profit or even non-profit endeavors. If you plan to take the Part 107 exam, you can find plenty of help and tips in this book. However, my goal is not to help you buzz the test; my goal is to help you understand and apply what the FAA wants you to know about drone piloting.

Foolish Assumptions

I wrote this book for you, the reader, because frankly, I needed this book some time back. I jumped into drone piloting head first, and then obtained my Part 107 exam certification, and *then* I opened a small business. A book like this would have saved me an immense amount of time and trouble figuring things out as I went along. So every bit of this book is dedicated to identifying and explaining what would have helped me (and now you) the most in my drone piloting journey.

In this book, I make no assumptions about your current knowledge of drone piloting, or even of how drones work. Maybe you're thinking about buying your first drone, or you've just bought your first drone, or maybe you've been flying for years. My only assumption is that you want to understand drones and take your drone piloting to the next level, and this book can help you do just that.

So, I wrote this book to help you fly higher . . . and faster . . . and to get your drone piloting goals off the ground quickly!

REMEMBER

This book isn't about a particular drone brand or model. You can use *Drone Piloting For Dummies* with whatever drone you own. This book also isn't a test prep book for the Part 107 exam, although I do explore a lot of the type of content you can expect on the test. The FAA expects you to know this stuff regardless of whether you decide to get certified.

Icons Used in This Book

Icons appear here and there in the book's margins to alert you to bits of information to pay special attention to, as follows:

TIP

A tip is meant to enhance your understanding or give you an additional idea to consider.

WARNING

When you fly a drone, you're flying a machine that could potentially harm a person or property. If something could potentially harm your drone, you, another person, or someone's property, this icon gives you a heads-up.

REMEMBER

Keep this information in mind at all times!

Beyond the Book

Thanks to my hard-working good friends at Wiley, extra content accompanies this book. Just visit www.dummies.com and search for *Drone Piloting For Dummies* to find the following:

>> **Cheat Sheet:** I've created a Cheat Sheet that offers quick tips for buying a new drone, flying safely, and starting a drone piloting freelance business.

>> **Updates to this book:** If this book needs updates, you can find them here as well.

Where to Go from Here

Each chapter serves as a reference for a specific topic in the drone piloting world. Each chapter is self-contained, so you don't have to read this book in order. For times when you need to refer to content from another chapter, I direct you to where you need to go.

With that said, you may find it best to read at least some of the chapters in order because the information in some will make more sense if you've read the previous chapters in a part. This situation applies especially to Parts 2 and 3.

TIP

Take your time and have fun along the way. And don't forget to fly! The more you practice, the better your skills develop, so get out there!

1

Becoming a Drone Pilot

IN THIS PART . . .

Explore drone piloting jobs and career opportunities.

Get familiar with your drone and flying it.

Practice drone flight techniques and maneuvers.

Take care and manage your drone and controller.

Chapter **1**

Finding Your Place in the Sky

As someone who loves nature photography, my purchase of a drone provided a natural extension of seeing the world around me. I remember nervously taking off for the first time, rising into the sky, and seeing the world from a completely different point of view. "I'm flying!" I thought. And that was it; I was hooked!

That first drone flight led to more drone flights, shooting photos and video in many different locations (of course I take my drone on vacation!). Those flights led to a new drone, and then Federal Aviation Administration (FAA) certification, and then another drone, and then people starting to ask, "Could you take some photos of our house that's under construction?" and so forth.

Most people sort of fall into a drone piloting career. You start out with a hobby, and the more your hobby grows, the more you may start to think, "Maybe drone piloting could be a career for me." Whether you're a committed hobbyist or considering expanding your horizons, this book covers all the essential aspects of drone piloting. From choosing a drone, learning to fly it, studying for the FAA certification exam, and understanding flight rules and safety to pursuing a drone piloting career, you can explore everything you need to know to find your place in the sky.

In this first chapter, you get an overview of drone piloting as both a hobby and a profession. You also delve into how drones work and what you need to know about shopping for and buying a new drone.

Exploring Drone Piloting

As is true of most popular technology, drones, or Unmanned Aircraft Systems (UAS), have changed a lot in recent years. Frankly, drones are really good these days. They fly well, have a lot of built-in technology rich with safety features, and their cameras are fantastic. All kinds of drones are designed for different industries and with different purposes in mind. You can buy a drone that will actually fly and take photos and video for under $50. Or, you can spend tens of thousands of dollars for a wide variety of sophisticated drone equipment. The sky is truly the limit, and the drone industry is vast and complex.

Piloting as a hobby

Years ago, my wife said, "You like photography; why not get a drone?" Honestly, I didn't think much of it at the time. It seemed that every extra dollar always went to some kind of camera or lens for photography at ground level. In fact, I even had a water camera to get some great snorkeling shots. I just didn't think that much about drones.

But as time went on, I discovered that many drones have very good cameras, and the angles I could capture could give me an entirely new look at the world. So I bought an inexpensive drone and played around. And it was fun. Something about the experience of flying is thrilling. Later, I purchased a prosumer-level drone (a cross between a professional- and consumer-level drone) and started getting serious about capturing great photos and videos. I even took this drone on vacation and shot the details of a coastline cliff. I remember saying to myself, "You would have never captured this footage without the drone."

TIP

My point is that I started out at the hobby level for fun. The odds are good that you started in the same place or are about to start there. For many people, though, this hobby tends to lead to professional considerations.

Piloting as a profession

For me, the turn from hobbyist to professional work began with a simple conversation with a friend who is a building contractor. He said, "You know, it would be great to have some aerial footage of some of my building projects. I could really use those on my website and social media." That conversation moved me in the direction of taking a hobby to a more professional level. Today, I shoot content for real estate agents, contractors, landscape designers, and a host of individuals who need aerial footage of personal property or events. I enjoy photography, so I stay in that realm with my drone.

However, your story may be different. You may be interested in piloting a drone in a completely different career path. Or you may already work in an industry that uses drones, so it's natural to think, "If I could pilot in my existing industry, I could advance my career!" That's not a bad thought.

TIP

In fact, drones are used by many different industries (see Figure 1-1), and many drones are even designed for specific industries as well. Here are just a few examples of industries utilizing drone technology today:

>> **Construction:** All kinds of construction projects use drones to get aerial views — from homes, commercial buildings, roads, bridges, and much more.

>> **Real estate and marketing:** Drones are commonly used to shoot aerial content for real estate and related marketing needs.

>> **Delivery:** There is a growing trend and many new trial programs for using drones in package delivery and fulfilling e-commerce orders.

>> **Law enforcement and public safety:** Drones are becoming commonplace in law enforcement, traffic safety, and search and rescue.

>> **Agriculture:** Specific types of drones can be used for all phases of agricultural management.

>> **Security:** Security and surveillance operations often use drones.

>> **Film:** The commercial film industry uses drones and employs them in almost all movies filmed today. Social media influencers also use them frequently in content development.

Note: In case you're curious, the industry that uses drones the most is real estate. Customers basically expect aerial views of houses, neighborhoods, and properties, so drone pilots are frequently employed for real estate shoots.

TIP

If you want to know more details about some specific kinds of drones that are used in specific industries, check out Chapter 19 for some examples.

Andrey Popov/Adobe Stock Photos

Shutter2U/Adobe Stock Photos

FIGURE 1-1:
Drones are used in many different industries.

kinwun/Adobe Stock Photos

Should you become a remote pilot?

If you're a drone hobbyist, you may reach that point where you wonder, "Should I become a professional remote pilot?" Of course, only you can answer that question, but before you ever begin the process of moving toward professional piloting, here are a few things to consider:

>> **Do you see opportunities?** As you think about professional drone piloting, do you see opportunities near you? Do you already know some people who would hire you for some projects? Do you work in an industry that uses drones and do you see a possibility of career advancement?

>> **Do you like working with people?** Here's a reality: You may enjoy flying your drone for fun and personal enjoyment, but flying professionally means you'll fly missions for someone else. That in turn means that you'll need to make other people happy with your work and fly missions that adhere to a set of goals defined by your customer. Does working under these conditions sound enjoyable to you?

>> **Are you willing to study?** In a nutshell, you must have an FAA Part 107 license to do virtually anything at all commercially. Are you willing to study and learn all the rules and regulations that the FAA expects you to know?

Along with the preceding questions, it's also important to take a hard look at yourself. Effective and safe drone pilots share some common characteristics. Do these sound like you?

- **Drone pilots are lifelong learners.** Effective drone pilots are always learning something new. The drone industry is about technology, and technology is always changing. Do you enjoy learning and reading about new drone technology, features, and related issues? Do you enjoy applying that new knowledge and putting it to work?

- **Drone pilots are methodical.** Safe drone pilots always think about safety and use checklists, employ effective methods of organization, and keep their drones in tip-top shape. Are you methodical in your thinking and your actions?

- **Drone pilots like to plan.** Good drone pilots carefully think through their missions, make notes, and execute plans. Do you like making plans, or are you more impulsive?

- **Drone pilots remain calm.** As a professional drone pilot, you may have to work with a frustrating client, deal with a frustrating mission, or grapple with things that go wrong during a mission. Good drone pilots remain calm because they know emotions can impact the safety of any mission.

- **Drone pilots have good communication skills.** Drone pilots may have to communicate with a wide variety of people, even during missions. Are you a clear and concise communicator?

- **Drone pilots pay attention to detail.** A common aviation term for drone piloting is *situational awareness*. This means that as a pilot, you're paying attention not only to your drone but also to the mission parameters and everything else that's happening in the area you are flying.

- **Drone pilots are technically proficient.** Safe and effective drone pilots are very good at maneuvering the aircraft and using the remote control (RC). You become technically proficient with a lot of practice. Do you like to fly a lot and constantly work to improve your skills?

Whether you wind up evolving from a hobbyist to a professional drone pilot comes down to your goals, skills, and drive. I would say the odds are pretty good that you're ready to start moving in the professional direction — after all, you're reading this book!

Looking at How Drones Work

Most people rely heavily on technology today, but fortunately, few of us need to understand how it works. For example, you probably have a cellphone in your hand a lot of the time, but do you really know much about its inner workings?

As with most technology, you don't have to understand many of the technical details that operate under the hood of a drone in order to fly one. Unless you want to design and build drones, you probably don't particularly care about *how* it works, as long as it works.

TIP

It's a good idea, however, to know some of the basics because knowing how a drone flies can help you be a better pilot and make better piloting decisions in a variety of flight situations. Specifically, you should know how drones fly, how the remote controller works, and how the gimbal helps you take great photos and videos. Don't worry; I skip the boring technical stuff and just tell you what you need to know.

How drones fly

Although there are many different kinds of drones, some of which even work like airplanes, most drones are effectively *quadcopters*, which refers to a type of drone that uses four rotors. Basically, a drone has four different sets of propellers on each of its four arms. These four rotors provide all the maneuverability you get when you fly the drone.

To understand how these rotors work together, you need to get familiar with a few terms and what these mean for your drone's ability to fly.

Thrust and drag

Thrust is the force that is necessary for your drone to fly. In other words, the rotors have to spin fast enough to lift the drone's weight up into the air. While that's happening, the opposite force, called drag, is at work. *Drag* is created by friction in the air as well as air density. The drone must have enough force to overcome drag or it can't fly. This is why your drone will perform better or worse depending on how much drag exists from the air pressure and other atmospheric conditions.

Lift

Each of the four rotors has its own motor and set of propellers. The moving propellers push against the force of gravity, giving the drone *lift* to fly vertically off the ground. To ascend (fly up), the rotors spin faster. To descend, the rotors slow down. Both ascending and descending are functions of lift.

Torque

Torque is the rotational force that each rotor creates. You can think of it this way: The torque that is created moves the drone in a direction. However, in order to control the drone's movement, you need opposite forces that cancel the effect of torque. Because drones are quadcopters, each rotor has a diagonal opposite that spins in the other direction. This oppositional force cancels the torque so that the drone can hover in one position and fly in the direction you want, as shown in Figure 1-2.

FIGURE 1-2: Diagonal rotors cancel out torque.

Pitch

Pitch refers to the drone's ability to move forward or backward. When you want to move forward, the drone's back propellers speed up, pitching the drone's nose down a bit so that the drone flies forward. To fly backward, the reverse happens: The front propellers speed up, pitching the tail downward, and the drone flies backward. Figure 1-3 depicts the pitch for flying forward and backward.

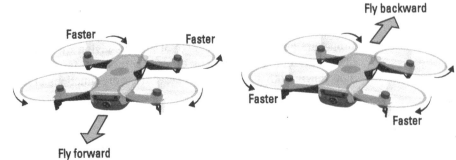

FIGURE 1-3:
Pitch forward
or backward.

Roll

Roll refers to the drone's capability to fly to the left or right. When you want to fly to the left, the right two propellers speed up, pushing the drone to the left. To fly rightward, the reverse happens: The left two propellers speed up, pushing the drone to the right. Figure 1-4 demonstrates flying to the left or right.

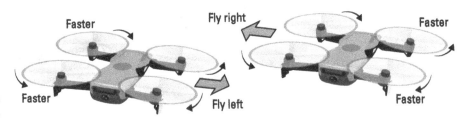

FIGURE 1-4:
Roll left or right.

Yaw

Yaw results from the use of torque to make the drone turn to the left or right. To invoke yaw, one of the diagonal sets of rotors slows down. To turn the other direction, the reverse happens. Yaw is easiest to understand with an image. Take a look at Figure 1-5.

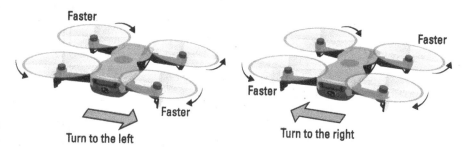

FIGURE 1-5:
Turn right or left.

Faster

Faster

Faster

Faster

Turn to the left

Turn to the right

Using the remote controller (RC)

You control the movements described in the previous section — lift, pitch, roll, and yaw — by using the two joysticks on the remote controller (RC); see Figure 1-6. Although some drones allow you to reconfigure the joysticks, virtually all drone joysticks work the same way, as follows:

>> **The left joystick controls lift and yaw.** Using the left joystick, push up to make the drone ascend. Push down to make the drone descend. Push left to make the drone yaw (turn) to the left, and push right to make the drone yaw (turn) to the right.

>> **The right joystick controls pitch and roll.** Push the joystick up to pitch forward or down to pitch backward. Push the joystick left to roll (fly) left or right to roll (fly) right.

TIP

If you're just learning to fly, it's a good idea to practice just these basic movements until you get comfortable with the joysticks. It's not unusual to get yaw and roll confused, but the more you practice, the more quickly the joystick movements will become second nature.

Of course, you're not limited to one movement at a time. You can use both joysticks together and fly with a lot of versatility. Consider a couple of examples. Say you want to fly forward (pitch), but at the same time slowly rotate the drone to the left (yaw). If you're shooting video, this combination of movements will give you a slow, panoramic shot. To move using pitch and yaw, you use the right joystick to move forward and the left joystick to rotate to the left, as shown in Figure 1-7.

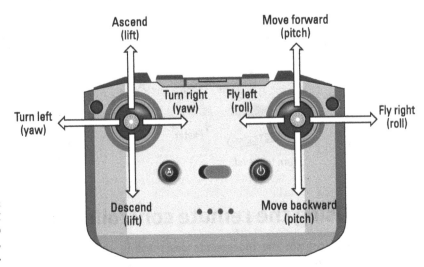

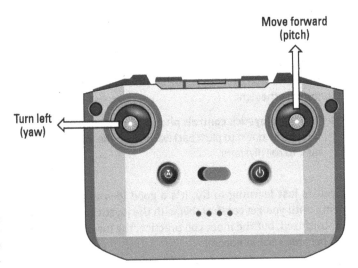

You can also use one joystick for a combination maneuver. Say you want to fly backward (pitch), but at the same time fly to the right (roll). In this case, you use the right joystick positioned between the down pitch and the right roll, as shown in Figure 1-8.

With the quadcopter and two joysticks, you can basically fly in any direction with any movement you want, which, when you think about it, is pretty amazing!

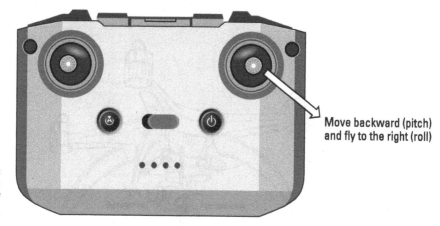

Move backward (pitch)
and fly to the right (roll)

FIGURE 1-8:
Using a
combination of
pitch and roll.

TIP

Want some good flying practice? Check out Chapter 3 for some different exercises that will help you practice the basic skills detailed in this section.

Understanding camera gimbals

Here's a quick but important note about the drone's camera. A good drone with a good camera has that camera mounted on a gimbal. A *gimbal* is a device that lets the camera move and remain stable, independent of the drone's movements. In simple terms, the motors in the gimbal keep the camera stable but also allow you to rotate it using the drone's RC. Without a gimbal, the camera would be subjected to the drone's movements and vibrations. You want a drone with a good gimbal, or you shouldn't expect to capture images or video that aren't blurry or jerky.

Gimbals are typically manufactured as two-axis or three-axis. Naturally, a three-axis gimbal does a better job because it moves on three axis points instead of two. A two-axis gimbal handles yaw and roll (turning and flying sideways), but it doesn't help you when you're flying forward and backward (pitch). A three-axis gimbal keeps your camera stable with all three. Now, that's a win!

TIP

If you're thinking about buying a new drone, don't forget the gimbal! Look for models that have a three-axis gimbal and you'll be much happier with your photos and videos. Figure 1-9 shows how the gimbal can move.

FIGURE 1-9:
A gimbal moves
to keep the
camera straight
and free of
vibration.

Shopping for a Drone

Shopping for a new drone can be overwhelming because you have so many to pick from with so many different features (and, of course, different price points). What drone do you really need, and how much money are you willing to spend? These great questions aren't always easy to answer. Choosing the right drone requires some thought and often a lot of reading and checking out reviews from other customers.

To make matters more confusing, you need to understand a big FAA rule before you shop at all. Read on.

Checking out the 250-gram weight maximum

If you're newer to the drone world and shopping around for a drone, you'll see drone ads that say things like "Under 250-grams!" or "Fully FAA compliant!" These ads speak to the FAA rule I mentioned previously.

REMEMBER

The FAA has determined that all drones weighing over 250 grams must be registered with the FAA. Registering is not that big of a deal; one part of it involves filling out an online form at https://faadronezone-access.faa.gov/#/, paying five dollars, and affixing a registration number to your drone. You also need to complete both your personal registration and the TRUST certification (see

Chapter 5 for details). With those requirements in mind, many drone manufacturers build small drones that weigh under 250 grams so that you can avoid registration.

Also, drones must have a remote ID, which is like a digital license plate, either installed or added to it. Most newer drones have a remote ID already built in, so an ad touting "FAA compliant" typically means that the remote ID is ready to go or the drone weighs under 250 grams (or both are true). Check the fine print, of course. Also, see Chapter 6 for more information about remote ID.

REMEMBER

The registration information presented here is for the United States. If you fly in another country, you need to check out the rules and procedures for that country. Fortunately, drone piloting rules tend to be very similar everywhere.

REMEMBER

Note that you can't use any drone for any kind of commercial purpose, whether you're paid or not, without a Part 107 license. You can find out all the details about Part 107 certification in Chapter 5.

Exploring drone categories

If you're drone shopping, it's helpful to think about categories of drones as you would different categories of automobiles. Drone categories differ in the drone's price points, features, and the purposes the drones serve.

In an attempt to help you make sense of the shocking number of drones for sale, I've sorted them into categories and included a few examples. Note that the categories are just a mental way to organize the drones, and there's cross-over between them, but hopefully the following sections can help you get a sense of what's out there. I also note some brands and a few models as examples of a drone category — not necessarily as recommendations.

Toy drones

Toy drones are basically just that: flying toys. They're small and easy to use, and most of them have software to download to your phone and use with the drone's RC. Typical pricing may range from $30 to $100. The Radclo drone pictured in Figure 1-10 is a good example. I paid $50 for this dron; it flies and it has a 1080p camera so that you can take some photos and videos with it.

However, don't expect much from drones in this category. (Considering that they're the cost a dinner out for two, you probably shouldn't.) One of the big complaints about toy drones is that they are difficult to control. They weigh so little that they can't handle any wind and tend to drift around while hovering because of poor calibration and stabilization. Also, they lack gimbals, so the photos and video you take are often blurry. These drones typically have no automated features, GPS, or safety sensors.

TIP

What I dislike about toy drones is that they're frustrating to fly and use, so someone's likely negative experience of this toy as their first drone may give up on drone piloting before they ever get started. That said, they are cheap, and they do fly; you just need to manage your expectations with drones in the toy category.

FIGURE 1-10:
The Radclo drone.

Photo: Curt Simmons

Consumer drones

Drones in the consumer space are a big step up from toy drones. They're small, with most weighing less than 250 grams, but they're larger and provide more features than most toy drones. Drones in the consumer category typically run from a couple hundred bucks to around $600.

Common drones in this category are the Holy Stone HS600, the DJI Mini series, the Ruko F11 series, the BWINE F7 series, and the Protensic Atom. What you get in this category varies by the drone.

The Holy Stone HS600, shown in Figure 1-11, is a good example of a mid-priced consumer drone. It provides a 4K camera with a two-axis gimbal and offers good wind resistance. The drone also uses GPS and can return to the home point automatically (see Chapter 2 for more information about this feature). It's easy to fly and gets good reviews.

TIP

Drones in the consumer category typically lack sensors and other advanced flight features. However, many of them are remote ID–compliant (an FAA requirement — see Chapter 6 for details). You may find the software and remote control features lacking with drones in this category, but they do work well and generally have decent cameras. Not bad, considering the price!

Photo courtesy of Holy Stone

Prosumer drones

The category of prosumer drones is a bit difficult to define. Some higher-end consumer drones may be considered prosumer, whereas some high-end prosumer drones start to slide into the professional category (covered in the next section). Prosumer drones generally seem to range in price from $700 to around $1,300 and are used by hobbyists just for fun, but they also serve ably for real estate photography, roof inspections, and more simple professional activities.

Drones in this category often fall just under the 250-gram weight limit; most are remote ID compliant and they have numerous features. Pricing can depend on what accessories you want to buy to go with the drone, including the number of batteries and the type of remote controller. (Some of these drones have built-in screens, so you don't have to use your phone, which is nice).

In the prosumer and professional market, DJI is the leader. In fact, DJI drones make up about 70 percent of all drones sold in these categories, and for good reason. They're just really good drones with good camera systems. The DJI Mini 3 / 4 are very popular drones in the prosumer category. Of course, competition is always good, so in the prosumer category, you'll find drones such as the DJI 3 / 4 series and the DJI Air series. Autel Robotics also competes in the prosumer category with the Evo Lite and Evo Nano drones. EXO Drones also competes in this space with the Blackhawk 3 (pictured in Figure 1-12).

Most drones in this category have the same basic features, such as 4K cameras with good sensors (which equals very good photos and videos), decent flight time, GPS, return-to-home features, automated flight modes, three-axis gimbals, and three-way collision avoidance sensors.

So how do you choose? I provide some guidance in the "Shopping Tips to Keep in Mind" section, later in this chapter.

FIGURE 1-12:
EXO Blackhawk 3.

Photo courtesy of EXO Drones

Professional drones

Professional-level drones are just that — drones people tend to use for professional purposes, professional-level photography and videography, and superior flight performance. These drones often cost several thousand dollars and can even soar upward of $10,000, depending on what you want and need for your profession.

DJI's flagship professional drone is the Mavic 3 Pro (pictured in Figure 1-13). I fly this drone as well as the DJI RC Pro (the professional remote controller), and both are fantastic. The Mavic 3 has plenty of features, flies great, and contains a Hasselblad triple camera system for amazing photography work.

Also in the professional category are drones from Parrot ANAFI as well as higher-end models from Autel Robotics, such as the Evo 2 Pro.

As you continue to climb the price ladder in this category, you might consider the DJI Inspire and the Autel Evo Max. These models are expensive, and at this point, you start moving into industry-specific drones.

TIPS FOR SIFTING THROUGH REVIEWS

If you're like me, you use online reviews to help you make purchases. However, reading through reviews can make you feel like you're sinking to the bottom of the ocean, and AI-generated reviews (which compile what a bunch of individual reviews said), often don't give you much detail to work with.

Here are a few tips. First, ignore the five-star reviews. I'm not saying they're not valid, but glowing reviews really don't help you decide. If the drone is great, then that's great, but what you really need to know are the negatives.

Second, look for themes. If one review says, "The propellers flew off and I crashed" but no one else says that, ignore it. Sometimes bad reviews happen because of a faulty drone, but sometimes they happen because of user error. Instead, look for themes. If numerous people say things like, "The drone loses connectivity with the controller," or "Tech support doesn't respond," those are warning signs because several people are having the same experience.

Finally, YouTube reviews can be good, but know that some YouTubers are given the product for free, which can make them biased, and sometimes they review products that they're already fans of. Also, some like to pick on aspects of a product that don't concern most people. So, again, if you use YouTube reviewers, watch several and look for common themes in what they say.

FIGURE 1-13:
The DJI Mavic 3
Pro drone.

Photo: Curt Simmons

Industry-specific drones

Industry-specific drones are designed for use in law enforcement, tunnel inspections, agriculture, and even the film industry. I show you a few models and include some details about them in Chapter 19.

Shopping tips to keep in mind

So with a review of some of the drones in different categories, you're likely closer to choosing a drone you want and need. To help you with that decision, here are few shopping tips to keep in mind:

>> **Pay attention to the camera and gimbal.** You want the best camera you can afford, and you'll be much happier with a three-axis gimbal. As you're shopping, pay close attention to those details first.

>> **Many drone companies use "flight time" as a selling point.** After all, longer flight times are good. However, I would basically ignore these advertising points. Although a particular drone may fly a bit longer than a competitor, your experience with flight time will vary wildly based on wind conditions, temperature, whether the air is dense or thin, and so forth. A long flight time means flying in perfect conditions, so unless you live somewhere perfect, your flight time per battery will vary.

>> **Think about accessories.** Sure, you need to focus on the drone, but as you're counting your hard-earned dollars, keep in mind that you may want an upgraded remote control and you probably want multiple batteries as well. As you shop, keep the accessories and extras you'll want or need in mind because these do drive up the cost.

>> **Spend what you can within reason.** A $900 drone is going to be much better than a $100 dollar drone. You tend to get what you pay for, so spend what you can and get the best drone you can.

>> **Check reviews about the drone company's tech support.** If you see numerous negative reviews about poor or nonresponsive tech support, I would move on.

>> **Finally, always keep in mind your goals.** Why are you buying a drone and how do you plan to use it? A firm understanding of your plans for the drone will help you make the best decision about which drone is right for you.

Considering Warranties and Insurance

Shopping for a drone is always exciting, and thinking about flying that new drone is even more thrilling. No one really enjoys thinking about warranties and insurance (at least until you crash into the lake), so this section offers a brief look at what you need to consider. (And don't skip this section!)

Basic and extended warranties

Most drones come with an included, basic warranty, such as a 90-day warranty. Basic warranties cover you for mechanical failure or a problem with the drone or controller. They don't cover damage from user mistakes.

TIP

Sometimes the manufacturer offers an extended warranty, which gives you the same protection for a longer period of time. However, such warranties don't cover damage from a crash (or if you drop it on the driveway). A drone is basically a flying camera, and accidents happen. Seriously consider other insurance plans. Read on.

Third-party plans

Typically, you can buy a warranty plan from a third party. For example, if you buy a drone from Amazon.com, you see an option to add a warranty from another company, such as Asurion or Allstate. Some of these options do include coverage for "drops, spills, cracked screens" and so forth, so they can be worthwhile. However, read the fine print carefully and know exactly what the insurance will cover and whether they have a deductible.

TIP

Some stores also offer their own warranty plans. For example, Best Buy's Total Tech covers everything you buy for an annual fee. These plans can be very helpful, but again, make sure that you know what you're getting and exactly what they'll cover.

Replacement plans

Many drone companies offer a replacement plan (often called a "Care" plan) that covers your drone in the event of a crash or what's called a *flyaway*. That's when something goes wrong with the drone software or hardware, you lose RC control, and the drone flies away into never, never land. The plan covers the cost of repairing the drone, or, if you need a replacement, you'll get one at a greatly reduced cost. Plans like this aren't cheap, and the cost is typically based on the cost of the drone, but they are good. It's always hard to throw a few extra hundred dollars in the mix when you're buying a drone, but I suggest you carefully consider purchasing a replacement plan. Maybe you won't need it, but if you do, well . . . then you do!

As with all plans, make sure you know what you're getting for your money.

Additional insurance

You may also be able to purchase additional insurance from larger companies such as State Farm, Allstate, Geico, and others. You can sometimes add your drone as a part of your homeowner's policy. However, this option varies a lot from state to state, and if the drone is covered under your homeowner's policy, it's probably still subject to the deductible (which may not really help you in that case). Also, some companies won't cover the drone if you take it out of the country, such as on a vacation.

The place to start when considering additional insurance is with your insurance agent. They should be able to answer these questions, or at least find out. That's their job, so don't hesitate to bug them about drone insurance options that may be available to you.

IN THIS CHAPTER

» **Registering your drone and getting ready to fly**

» **Understanding GPS and using B4UFLY**

» **Getting to know your drone's features and procedures**

» **Taking off for the first time**

Chapter **2**

Flying for the First Time

f you're just getting started with drone flying, you're in for a lot of fun. Perhaps you have purchased your first drone, and you also picked up this handy book to help you get started. Congratulations! There's nothing quite like flying your drone for the first time.

You're likely feeling a bit of anxiety, though. After all, you just spent some hard-earned money on a robot that can fly. No one wants to experience a drone crash the first time they head to the skies (or ever, for that matter).

The good news? Today's drones are highly advanced. They fly extremely well, and most have built-in safety features that help you consistently experience safe and enjoyable flights.

In this chapter, you find out all the vital information to register your drone with the Federal Aviation Administration (FAA), a necessity before flight for most drones. You also check out GPS and the FAA's B4UFLY apps. Finally, you explore your drone's features and procedures and review all you need to know to fly for the first time. Get ready to put your drone in the air and have some fun!

WARNING

Most first-flight accidents happen because pilots are not prepared. They didn't read the procedures and aren't familiar with the remote controller and software before leaving the ground. You've waited this long, so slow down and gather knowledge before you fly so that you can have a positive first-flight experience. This chapter is here to help you do just that!

Registering Your Drone and Checking for Updates

Before you even think about flying your new drone, you must first understand what the FAA expects from you concerning registration. Also, before you fly, you'll want to make sure that your drone is updated with the latest software and firmware updates.

Registering your drone with the FAA

First things first: The FAA calls flying drones Unmanned Aircraft Systems (UAS). Everyone calls them drones for short, but just know that some drones are used on land or even underwater. This book refers to a UAS as a drone.

Next, FAA recognizes two basic categories of drone pilots: recreational pilots and certified remote pilots. It's important to understand these differences before you ever fly your drone.

Keep reading to find out the different drone registration rules for each type of pilot.

Recreational pilots

Recreational pilots fly for personal enjoyment and fun. You don't do anything commercially. In other words, you don't take drone real estate photos or even use your drone photos in monetized social media. You would say, "I just want to fly for fun and take photos and videos for my personal use." If that's you, here are the rules:

- **Fly only for recreational purposes (personal enjoyment).** To be a recreational pilot, you can't do anything with your drone that doesn't fall under the recreational category.

- **Follow all FAA safety guidelines.** In other words, the FAA expects you to know safety guidelines and procedures.

- **Keep your drone in your visual line of sight (VLOS).** As you fly, you must be able to physically see your drone in the sky. Another person (called a visual observer) is allowed to stand next to you and help you keep track of it.

- **Give right-of-way to all other aircraft.** You must yield the right of way to another other aircraft, including other drones. In other words, stay away from any other aircraft and let that aircraft pass.

>> **Fly at or below authorized altitudes in various airspaces.** The National Airspace System (NAS) provides information about different kinds of airspaces, and you must fly at or under the altitude requirements for those airspaces. See Chapter 9 for more information about NAS.

>> **Fly at or below 400 feet in uncontrolled (Class G) airspace.** This means that 400 feet is the maximum height you can fly in uncontrolled airspace. In controlled airspace, that maximum may be much lower, or you may not be able to fly at all. Controlled airspace is controlled by the FAA; uncontrolled airspace isn't. You can find out more about airspace in Chapter 9.

>> **Do not fly your drone in any manner that endangers the safety of the National Airspace System.** In all situations and locations, make sure you're flying safely.

You can find more about these foundational rules and check for any additional information at https://www.faa.gov/uas/recreational_flyers.

Along with the foundational rules, here are some specific rules about your drone's registration and the TRUST test for recreational flyers:

>> **You can fly without registration if your drone weighs no more than 0.55 pounds (250 grams).** There are many starter-level drones that weigh under the 250-gram threshold for this purpose. You get to fly a small drone for fun without registration with the FAA.

>> **If your drone weighs more than 250 grams, you must register it with the FAA.** If you need to register your drone, the process is quick and easy. Just go to the FAADroneZone website at https://faadronezone-access.faa.gov/. Create an account and then follow the prompts to register your drone. After you register, you'll receive a registration number for the drone that you'll need to affix to your drone using a permanent marker, an adhesive label, or another permanent method. This number will be your drone's registration number as long as you own the drone.

>> **Take the Recreational UAS Safety Test (TRUST).** The TRUST test is free, and it tests your knowledge of basic drone safety information. The FAA wants drone pilots to fly safely, so this test is designed to check your understanding of basic safety rules. After you pass the test, you should keep your TRUST certificate with you when you fly. To find out more and take the test, visit https://www.faa.gov/uas/getting_started.

Certified remote pilot

A certified remote pilot is someone who legally flies for commercial purposes (including nonprofit), government, small businesses, real estate — basically anything that's not recreational.

Certified remote pilots must follow the same rules as recreational pilots concerning safety within the National Airspace System (see the previous section).

Certified remote pilots must register all drones they use regardless of weight with the FAA at https://faadronezone-access.faa.gov/. You must also pass the FAA's remote pilot certification (Part 107). See Chapter 5 for more information about the Part 107 exam.

TIP

Many people are confused about the differences between recreational pilots and certified pilots. Here's the thing: If you do anything at all with your drone that is not strictly for personal, recreational purposes (whether you get paid or not), you need to become a certified remote pilot and pass the Part 107 exam.

Checking for updates

Your drone flies using hardware and software. As with computers and phones, the software that manages your drone's hardware (called *firmware*) is often updated; likewise with your drone's controller software. It's important to keep your drone's firmware and software up to date, so you should check for updates before you ever fly.

In most cases, you can check for updates using your drone's controller. You'll need to have your drone turned on as well. When you update, the updates are applied to both the drone itself and the remote controller (RC) software.

Usually, the update options will appear as a menu feature on the RC. Just check your drone's documentation for how to access that feature. In Figure 2-1, I'm checking for updates in a menu on the RC.

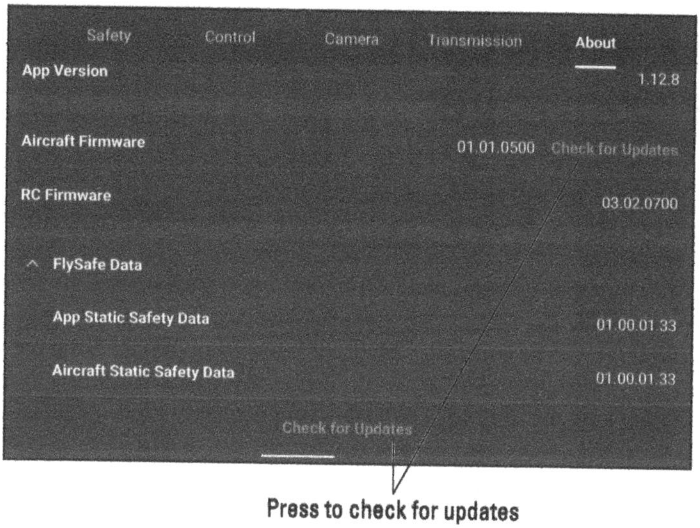

FIGURE 2-1:
Check for
firmware and
software updates
on the remote
controller.

Press to check for updates

TAKING YOUR DRONE ON AN AIRPLANE

Say you're going on that dream vacation and want to fly your drone while there. I've flown in lots of places around the world and, of course, it can be a lot of fun. But what about taking your drone on an airplane? This is often a point of confusion for people (and the internet offers a lot of incorrect information on the issue). It's perfectly fine to take your drone with you on an airplane as long as you follow the rules:

- **Drones and drone batteries need to fly in the cabin with you as carry-on luggage.** Never pack your drone, remote controller, or batteries in checked luggage — *ever*. Treat your drone as you would a laptop or phone. Take it on the plane with you.

- **Your drone's battery must be removed from the drone and carried separately.** Ideally, you'll put it in a fire-safe lithium-ion polymer (LiPo) battery bag, which you can buy from Amazon.com or anywhere that sells drones. Some airlines have restrictions on the number of drone batteries you can carry with you, so be sure to check with your airline for details.

As a side note, if you travel out of the United States, the same general rules apply. I've never had any problems with airport security or taking a drone on an international flight. Just check the airline you're flying for any specific details or rules.

However, if you travel out of the United States, you may find different rules for actually *flying* your drone at your destination, or you may need permission to fly it. Sometimes finding the right information can be sketchy, so do your homework with some good internet searches and make sure that you apply for permission if you need to do so.

Getting Familiar with Your Remote Controller

You use your drone's remote controller (RC) to control and manage your drone when you're flying it. There are two kinds of RCs: one that uses software you install on your phone and uses your phone as a screen and part of the controller; and one that is completely self-contained, with a built-in screen. If you're using a First Person View (FPV) system, the headset is the screen. As you can see in Figure 2-2, the controller on the left uses a phone, and the controller on the right has a built-in screen. RCs with built-in screens are much more convenient than using a phone, but not all drones support these kinds of RCs.

FIGURE 2-2: Two kinds of drone remote controllers.

Mario Arango/Getty Images

happyphoton/Getty Images

Before you fly, you'll want to get familiar with the different icons and control options you see on the controller. After all, that's how you'll manage and fly your drone. Different drone RCs provide different options for the controller and what you see on the screen. Most drones provide two joysticks so that you can control the elevation, direction, and movement of the drone. See Chapter 1 for details about the joystick controls.

For the RC software, you need to check out your drone's documentation for details about the software features (and downloading the software to your phone, if you're using that kind of RC). However, the good news is that most RC software provides the same basic features. Some common onscreen features you see are shown in Figure 2-3 and described in the following list.

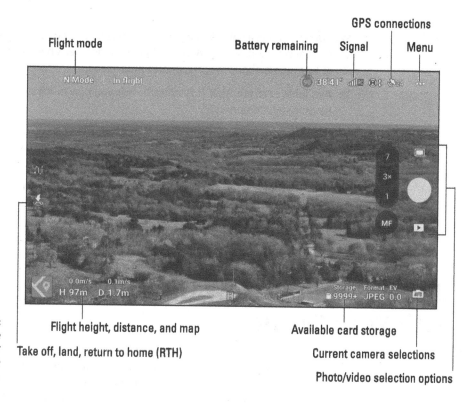

Flight mode

GPS connections

Battery remaining   Signal   Menu

FIGURE 2-3:
Drone remote
controller
software
features.

Flight height, distance, and map

Take off, land, return to home (RTH)

Available card storage

Current camera selections

Photo/video selection options

WARNING

>> **Flight mode:** Your drone may support different modes, such as Cine (slow), Normal, and Sport (Fast). Your controller shows you which mode you're currently flying, and you can switch modes in-flight.

Many drones have a Sport (also called Speed) mode that turns off the sensors, leaving you with no collision avoidance options. Use extreme caution when flying in Sport or Speed mode. Many accidents happen using this mode!

>> **Battery remaining:** This indicator tells you how much time you have left on the battery, often displayed in minutes and seconds.

>> **Signal:** This indicator shows you the radio signal strength between the RC and drone.

>> **GPS Connections:** This indicator shows you how many GPS satellite connections the drone has. See the next section to learn more about GPS and your drone.

>> **Menu:** Press this icon to access various menu and setting options.

>> **Photo/video selection options:** This group allows you to take photos and videos, and choose automated flight paths for video purposes. The options you have will vary based on your drone's capabilities.

>> **Current camera selections:** This group of camera icons shows you the current selection, such as Auto or Manual modes. See Chapters 17 and 18 to learn more about shooting photos and videos.

>> **Available card storage:** This indicator tells you how much storage space you have on the memory card.

>> **Flight height, distance, and map:** These indicators tell you the drone's current height and distance. Also, on some drones, you can switch to a map view.

>> **Take off, land, return to home (RTH):** The button option here allows you to take off or land the drone. You also choose an automatic Return to Home function (RTH). See the next section to learn more about RTH.

Understanding How GPS Works with Your Drone

You don't need to be an expert in the Global Positioning System (GPS) to use it. After all, you probably already use it on your phone and in your car. GPS is an amazing system that helps you know where you are in the world and how to get from one place to another.

The same is true for your drone. Your drone uses GPS to know where it is and where it took off. Without GPS, your drone flies "blind" and doesn't know where it is in the world. With GPS, you can see where the drone is flying via the video screen, but your RC may have a map view that you can use as well.

GPS ACCURACY

It's important to note that GPS is only accurate to around ten feet. This means that your drone will return home and land automatically, but it's unlikely to land in the exact spot where you took off. In my experience, it's usually off by a few feet. That's no big deal, but if you use a launch or landing pad, the drone will probably miss it by a foot or so when it lands. Landing pads are helpful in tall grass, sand, pebbles, and other locations where taking off and landing can throw debris into the drone's propellers. So, if you use RTH, just remember that you may want to cancel RTH as the drone gets close to landing, and then land manually so that you can adjust the drone's location in order for it to land on the pad. Landing pads are inexpensive and very helpful. You can find many options anywhere drones and drone supplies are sold.

Here's the skinny: When you get ready to fly, you put your drone in the location from which you want it to take off. Then you turn on the drone and the RC. The RC will locate GPS satellites and connect with that system. When you have enough satellite connections (typically, 14 or more is ideal), your RC will tell you that you're ready to take off.

Here's why it matters: Your drone probably has a feature called Return to Home (RTH). Using GPS, your drone can automatically come back to the place from which it took off and land by itself with the click of a button on the RC. Also, on many drones, if the battery power gets too low or the drone loses radio connectivity with the RC, the RTH features will take over automatically, enabling the drone to come home and land before it runs out of battery and crashes — a great feature!

When you or the RC initiates an RTH, most drones will fly vertically to gain elevation to a certain height (which you can configure on the RC). This way the drone avoids trees and other obstacles. Then, it will fly at this elevation to the return to home point. When it reaches the home point, the drone descends and lands

Downloading a B4UFLY App

The FAA has a program called B4UFLY. This service helps you know where you can and can't fly and whether there are any restrictions you need to know about. The FAA partners with different app and application providers, so you can choose an app and download it to your phone or computer, and then use that app when you're at a location to see what's legal and what's not. These apps are easy, helpful, and free.

For example, in Figure 2-4, I'm in Arches National Park in Moab, Utah. I'd like to fly my drone here and get some great photos. So, I check the app I'm using and find out that I'm in uncontrolled, Class G airspace (good), but I'm in a national park, and drone flights in almost all national parks in the United States are not legal (bad). So, I find out I can't legally fly here. However, I can get outside the park and check the app again to have it update the flight information for me. As you can see, the app can be very helpful to find out the rules for flying in a particular location.

TIP

You can find out more about B4UFLY and see what apps are currently available for download at https://www.faa.gov/uas/getting_started/b4ufly.

FIGURE 2-4:
B4UFLY apps help
you know the
flight rules for a
particular
location.

Using Your Drone's Crash Safety Features

Many starter drones (often called "toys") are not outfitted with exterior sensors. Without sensors, the drone doesn't know what's around it and gives you no crash protection. If you spend a bit more and get to the next level of drones that may be used for recreation or even commercial work, you'll probably get some sensors on the drone, and thus some crash safety features. For example, the DJI Mini series has some sensors and crash avoidance features.

Of course, RTH is a sort of crash protection feature in that the drone can come home by itself and land if the battery power is running low or if it loses communication with the RC. But what about avoiding collisions with trees and other objects? This is where sensors come into play, and they are very helpful.

Most drones that have sensors will give you a couple of obstacle-avoidance behaviors, among which you can choose as a menu option on the RC (shown in Figure 2-5):

>> **Brake:** This feature works just as its name sounds. If you fly your drone toward an object, such as a tree, the drone will automatically brake, or stop when it gets close to the object. You can't override this braking feature with

joysticks. The drone will just stop flying and hover until you change direction. For example, in Figure 2-6, the drone has applied a braking procedure to prevent running into a tree. Braking lines appear on the screen, showing the potential impact locations.

>> **Bypass:** Using the bypass feature, the drone will try to bypass and automatically fly around obstacles. For example, you can fly your drone through the woods and the bypass system will take over, maneuvering the drone automatically around the trees. Very cool!

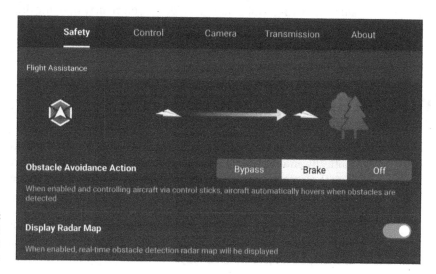

FIGURE 2-5:
Choose crash safety features on the RC menu.

Braking lines appear when the drone is too close to an obstacle

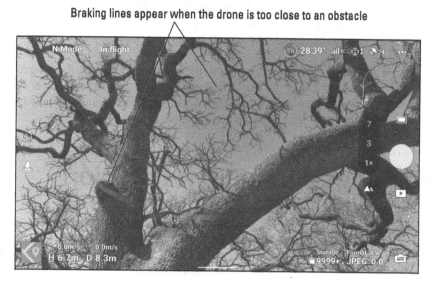

FIGURE 2-6:
Braking stops the drone from flying into an obstacle.

WARNING

Although the sensor system works well on drones, the sensors don't always see everything, especially small objects. Often, these sensors have trouble seeing small twigs or small branches on a tree and can run into those, causing a crash. (I'm speaking from experience here!) No sensor is as good as your own eyes, so always use caution and good judgment when you fly.

TIP

Make sure you understand your drone's sensor system, or lack thereof. Often, less expensive drones will provide front, rear, and bottom sensors but no side sensors. So, you have no collision avoidance when flying sideways. More expensive drones typically provide sensors in all directions. On the RC, you may be able to disable sideways flying, which is a good idea if you're new to flying drones and you don't want to chance running into something when you fly sideways. Check your drone documentation for details.

Understanding Geofencing

Geofencing is an automated procedure that some drone manufacturers use to automatically enforce no-fly zones or restrictions. The technology works with GPS and digital airspace charts. Instead of just advising you about the restrictions, the drone will enforce them and not fly.

You can sometimes override geofencing by entering information about the drone and the pilot, depending on the location. This authorization is usually temporary and may be limited in other ways.

Geofencing is a hotly debated issue. Those in favor say it helps enforce fly-zone restrictions, making the air spaces safer. Those against say it can hamper time-sensitive missions, such as search and rescue.

Regardless of your position, the use of geofencing is manufacturer specific, so you might check out the details with your drone manufacturer for a clearer understanding of how geofencing works with your drone and whether you can apply to override the setting in certain cases.

Choosing a First Flight Location

The more you fly, the better you'll be able to take off and land in tricky locations. Don't try one of those locations for your first flight. You want a good location that's wide open and easy for you to take off and land. Here are some quick tips as you think about your first flight location:

>> **Choose wide-open spaces.** That's not just a country song; it's good advice for the first-time drone pilot! Choose a location that's open and free from low-level obstacles. Avoid locations with a lot of people, cars, and houses. Some parks, sports fields, or remote locations away from everything are ideal for your first flight. A good question to ask yourself: If I crash, who or what would I possibly damage in this location? Let the answer to that question be your guide.

>> **Look for power lines.** Power lines are so common that we often ignore them, but your drone's sensors may not see them, and many people have inadvertently crashed their drones into a power line. When you choose a location, look up. Make sure there is nothing overhead that you might crash into.

>> **Choose a level place.** Don't take off on the side of a hill. Even what appears to be a gentle decline can cause you liftoff problems. You need a level spot, so be sure to look for one.

>> **Consider whether you need a launch pad.** When your drone takes off and lands, it can throw dirt, sand, pebbles, and other loose debris all over your drone. Tall grass can get tangled up in the propellers as well. For this reason, you may want to use a launch pad. They're inexpensive and available anywhere drones are sold.

Following Pre-Flight Checks and Procedures

All right, you've done all the preliminary work! You're familiar with your RC and you've run firmware and software updates. Now you're just about ready to go, but before you fly for the first time, here are some things to check:

>> **Batteries:** Make sure the drone's batteries and RC batteries (and your phone, if necessary) are completely charged.

>> **B4UFLY:** Use a B4UFLY app to check the location you want to fly. As a first flight, you should find a location that is completely unrestricted (Class G airspace).

>> **Weather:** Make sure the weather forecast is clear of any precipitation, and take a good look at the wind. Here's the thing: Wind is not your friend! Different drones can handle different wind conditions based on the drone's size and weight, but know that wind makes your drone hard to control and will greatly reduce the life of the drone's battery. If you fly in windy conditions,

make sure you're flying in an open area, and keep the drone close to you. You may have to land much sooner than you would expect if conditions worsen or if the battery levels drop quickly (and remember that you can keep a watch on the drone battery life on your RC).

>> **Propellers:** Unfold the drone's propellers (if necessary) and check them out. Make sure the propellers spin and are free of nicks and other damage. If there's a problem, see your drone's documentation for details about installing new propellers. See Chapter 4 for more details on inspecting and changing propellers.

>> **Crash safety:** Make sure a crash safety feature is turned on using the RC. Choose either Brake or Bypass, but use one of them!

>> **Carry your TRUST certificate or Part 107 license:** The FAA requires you to have these items with you when you fly.

>> **Check the SD card:** Make sure the SD card is installed correctly on the drone. You'll need it to capture photos and videos.

>> **Check surroundings for a visual line of sight (VLOS):** The FAA requires that you maintain a VLOS with your drone when you fly. Can you do that in your chosen flight location? Are buildings, trees, and other obstacles going to hinder your VLOS? If so, you might consider a different location to fly.

Taking Off and Landing for the First Time

The moment has come: It's time to fly! Get ready for a thrilling experience. Take your drone and RC to the spot you've chosen for takeoff, remove the lens cover, and then follow these steps:

1. **Turn on your drone and RC (connecting the RC to your phone if necessary).**

2. **Give the RC a couple of moments to connect to satellites.**

   The RC will probably tell you when you're ready to take off. Usually, you'll see a message that the home point is updated.

3. **If you see a message on the screen that calibration is necessary, follow the on-screen instructions to do that.**

4. **Choose a flight mode, such as Cine (slow) or Normal.**

   I strongly recommend that beginners fly in a slow mode or normal mode.

5. **When you're ready, stand 8–10 feet away from your drone and press the take off button.**

   Your drone will launch a few feet into the air and hover, as shown in Figure 2-7. Although unlikely, errors do happen, and you don't want your drone flying into you. So, stay back during takeoff. It's also not a bad idea to position the drone so that it's facing away from you when you launch.

Huseyin Bostanci/Getty Images

FIGURE 2-7:
Stay back when
you take off.

6. **Use the joystick (typically the right one) on the RC fly straight up.**

   I recommend flying to 200 feet (about 60 meters) or so. Doing so will get you above trees or buildings so that you can practice maneuvering and flying your drone. Remember to maintain a VLOS.

7. **Keep a watch on your remaining battery time on the screen as you fly.**

8. **When you're ready to land, fly back over the landing location and hover over it, and then use the RC to descend.**

   The drone will come straight down and stop a few feet from the ground.

9. **Continue holding down the joystick and the drone will slowly descend, land, and stop running.**

   If landing makes you nervous, remember that you can just press the Return to Home (RTH) button on the controller and the drone will do the work for you.

Whew! That's it — you did it. The more you take off and land, the easier it will be. And the more you practice flying, the more confidence you'll gain. Be sure to check out Chapter 3 for more details about maneuvering your drone.

What to Do When You Don't Know What to Do

It's not unusual in the early days of drone piloting to get a little nervous, distracted, or confused. Maybe you lose sight of your drone and you can't figure out where you are (things look different from the sky). Maybe you're flying and the wind gets too strong, or perhaps you see errors on the RC. Whatever the case may be, you have two simple actions to take:

>> **Stop:** Stop moving the joysticks. Your drone will stop flying wherever it is and just hover. This gives you a moment to collect your thoughts. Never keep flying in an erratic pattern without a plan. Just stop and hover until you know what you need to do next.

>> **Use RTH:** When you're not sure what to do, just have the drone come home. You can use RTH to get the drone back to you safely, and in the end, that's what you want.

Following Post-Flight Checks

When you finish flying, you should perform a few post-flight checks and activities. Get in the habit of doing these every time!

>> **Remove the battery from the drone.** You should not store your drone with the battery installed, so just get in the habit of taking it out every time.

>> **Charge the batteries and RC.** You want your drone ready to go at a moment's notice, so when you're done flying, charge everything so that you're ready for the next time.

>> **Check the drone for damage and clean it.** Inspect the propellers for any damage, and gently wipe down the drone and camera lens with a microfiber cloth. Keeping your drone clean keeps things working the way you want.

>> **Check the drone's batteries.** Your drone most likely uses lithium-ion polymer (LiPo) batteries. They will probably last a long time, but always check them. LiPo batteries can catch on fire, and swelling is a tell-tale sign. If the battery seems swollen or misshapen, you'll need to get rid of it and buy a new one. (No fun, but that's better than a fire.) If you need to get rid of a LiPo battery, discharge it if possible and then see if a recycling service near you takes them.

>> **Fold up the drone and store everything indoors and in a safe place.** This advice may seem obvious, but when you're not flying, fold up the drone and store it, the batteries, the RC, and anything else in a safe location indoors. Your drone may look cool sitting on your office desk, but it only takes one child — or cat — to knock it to the floor and cause substantial damage. Your drone is an investment. Treat it with care, and it will last you a long time.

DRONE CASES AND BACKPACKS

Your drone may have come with a case, but often the ones that come with the product are not great. It's better to invest in a good hard case or backpack designed for drones. In fact, you may find hard cases and hard-shell backpacks that are designed specifically for your drone's make and model. These cases and backpacks are often shockproof and waterproof, and they're well worth the money. One time, I wanted to fly in a mountain area that required an ATV to access it. I had my drone in a shockproof case, and as I was going up the mountain, the drone slid out of the ATV and onto the side of the mountain. Upon opening the case, I found no damage at all. That case was well worth the money! You can shop for cases and backpacks anywhere drones are sold.

Chapter **3**

Maneuvering and Flying Your Drone

A drone pilot flies a drone. That's pretty basic information, but the truth is that flying a drone is anything but basic. Your piloting skills are what will make all the difference. Whether you focus your attention on photography or on work toward more complex, industrial drone applications, your ability to pilot a drone is what will open the world of drone piloting jobs and careers for you. After all, drone pilots fly!

As with any skill, good drone piloting doesn't happen overnight. You have to develop and grow your piloting skills along the way. The more you fly and the more you practice a variety of flight skills in a variety of situations, the better pilot you'll be.

In this chapter, you explore all kinds of flight maneuvers, tips, and tricks. You find out how to perform foundational flight patterns as well as use built-in flight routines that your drone may have available.

TIP

Think of this chapter as a workshop. As you explore different flight techniques and maneuvers, you might want to stop reading periodically and practice flying for a bit. Again, the more you practice, the better you'll be, so charge those drone batteries and get ready to have some fun!

Exploring Foundational Flight Principles

Before you delve into the details of different flight maneuvers, it's a good idea to grasp some foundational principles. After all, you may know the drone's RC controls and how to physically fly the drone, but your piloting skills also hinge on your thinking and your attitude toward piloting.

The next several sections describe three foundational flight principles to always keep in mind:

» Safety first

» Consider the weather

» Always do pre-flight checks

Safety first

The worst thing that can happen as a drone pilot is crashing your drone into a person and injuring them. That's a terrible scenario to deal with, and the truth is these kinds of accidents can most often be avoided. They happen because too many drone pilots are thinking about drones and not people.

In one of the drone piloting social media groups I belong to, a group member posted a video of his drone flying over Athens, Greece. It was a cool video, but when someone noted that it's not legal to fly there, the group member responded, "I just fly where I want."

This is wrong thinking and the wrong attitude. Although no-fly zones and restrictions can be frustrating, they exist for a purpose. Often, that purpose is to prevent injury to people. In crowded areas, should you have an accident, the odds of you hurting someone are high.

TIP

A key foundational flight principle is to always think about safety before you think about anything else. The more you fly, the easier it is to become lax and comfortable. Mentally, fight that battle and strive to stay sharp. Always ask the question, "Am I putting someone in danger with this flight?" See Chapter 16 for more details about safety principles and procedures.

Consider the weather

The weather has everything to do with your flight and your ability to control and manage your drone in the air. When I say "weather," I'm not just talking precipitation. I'm also talking about wind. Drone pilots often underestimate how the wind speed will affect a flight and the drone's battery time. Your drone has to work much harder to fly in windy conditions, which means that you burn through the battery faster. So the second foundational principle of drone piloting is that to always consider the weather when you're getting ready to fly.

TIP

There are many types of drones, and they all handle wind speeds differently, according to their size and weight. Check your drone's documentation for details about flying in various wind speeds. Also, find out more about dealing with weather conditions in Chapter 14.

Always perform pre-flight checks

The more you fly, the easier it is to fly, but the easier it is to get complacent with pre-flight checks, too. Work on becoming a machine when it comes to your pre-flight routine! Before you fly, follow the exact same procedure every time as you perform (and never skip) all the pre-flight checks. These checks help to ensure that you have a safe and productive flight.

Did you check a B4UFLY app to make sure you're cleared for takeoff? Have you inspected the drone and propellers for damage? Are you sure the takeoff area is safe? Do you have the collision avoidance features turned on? These are just some of the pre-flight checks you should perform — every single time!

TIP

You can find more information about B4UFLY apps as well as a pre- and post-flight checklist in Chapter 2.

Practicing Basic Drone Flight Maneuvers

If you're new to drone piloting, there are a few basic flight maneuvers you should practice. The more you practice these maneuvers, the better you'll perform them, and they'll become second nature over time.

Most drones use two joysticks to control the movements of the drone, and a certain amount of muscle memory sets in as you practice. The more you fly, the more your muscle memory takes over, and you don't have to think through every action needed to make the drone do what you want. As with all skills, practice is key, so be sure to get a lot of flight time working on the basic drone maneuvers described in this section.

TIP

It's not unusual to get confused by the joystick controls as you fly. That's normal if you're new to drone piloting. Just take your time as you practice, and see Chapter 1 for more information about using the remote controller's joysticks.

WARNING

Fly in Normal mode as you practice, not Sport or Speed mode. If your drone supports a slower mode, such as Cine, use that.

Straight up, straight down

A good initial practice is to simply fly straight up and then come straight back down for a landing (see Figure 3-1). This exercise helps you work on keeping the drone in the same place with a vertical ascent and descent. To practice flying straight up and down, follow these steps:

1. **Take off and fly straight up to about 350 feet. Hover for a moment and take in the view.**

2. **Descend, flying straight down, and try to land in the same place you took off.**

   Repeat steps 1 and 2 a few times for practice.

3. **Fly straight up to about 350 feet; then descend and prepare for landing, but this time, try to land about six feet or so from where you took off.**

   This step helps you practice adjusting the drone's vertical descent and landing the drone where you want. This is a great skill because in many situations, you may take off in one spot but need to land in another.

TIP

Say you take off, ascend, and then descend to land. You'd think that the drone would land exactly where you took off. After all, you just went up and down. However, you notice that it's not in the same place when you land. This typically happens because wind has affected your drone's location and pushed it around a bit. Just remember that although you are controlling your drone, the forces of nature are acting on it as well!

TIP

Are you getting confused about which joystick does what as you fly? That's normal! See Chapter 1 for a detailed explanation of your drone's joystick controls.

FIGURE 3-1:
The straight up,
straight down
maneuver.

Straight forward, straight back

The straight forward, straight back maneuver (see Figure 3-2) helps you practice flying toward an object and away from it, all the while keeping that object in the center of your screen.

Follow these steps to practice the straight forward, straight back maneuver:

1. **Ascend to a good height that puts you above any obstacles.**

2. **Find any object to fly toward.**

   You want to put something in the very center of your camera frame, such as a distant tree or building.

3. **Without gaining or losing altitude, fly directly toward the object, keeping it in the center of the camera frame.**

   You may have to adjust your direction a bit as you fly — that's normal.

4. **When you get close to the object, stop and hover.**

5. **Fly in reverse away from the object, keeping the object in the center of the screen as you fly backward.**

6. **Repeat these steps several times.**

FIGURE 3-2:
The fly forward,
straight back
maneuver.

Circle

The circle maneuver is just as it sounds (see Figure 3-3). Fly your drone to about ten feet off the ground and practice flying in a small circle. This maneuver is all about joystick control, and many new drone pilots find it frustrating. The circle pattern tends to drift and get larger as you make each loop, so you have to really practice keeping the drone in the same location as you form the circle.

When you feel that you have the maneuver down, try reversing the pattern and create a circle flying in the opposite direction.

As an alternative, you can try flying in a circle around a stationary object, such as a flagpole. This physical item in the center of your circle can help you practice controlling the movements and distance (and keep your circle from becoming an oval).

WARNING

It's easy to crash into something with this maneuver if you're not in a wide-open area. Make sure that you fly in an open area, and with collision avoidance turned on!

Rectangle

The rectangle maneuver operates on the same principle as the circle, but you create a rectangle as you move (see Figure 3-4). Fly your drone to about 10 feet off the ground and try it. This maneuver is all about joystick control and flying hard lines so that you create right angles as you fly. As with the circle pattern, the rectangle tends to drift and get larger as you fly, so keep practicing.

When you feel adept at this maneuver, try reversing the pattern and create a rectangle flying in the opposite direction.

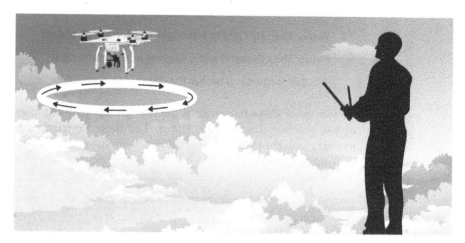

FIGURE 3-3:
The circle
maneuver.

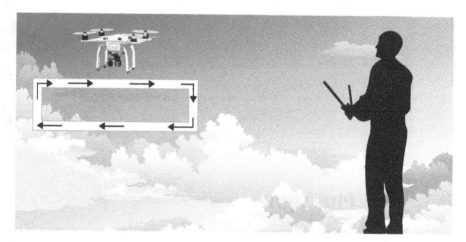

FIGURE 3-4:
The rectangle
maneuver.

As you practice, try modifying this maneuver to form a square rather than a rectangle. Also, try making the rectangle or square smaller as you practice. The smaller the rectangle or square, the more challenging!.

Triangle

The triangle maneuver is all about ascending and descending while moving horizontally at the same time (see Figure 3-5) to form an upside-down triangle. It takes some practice, and here's how you do it:

1. Take off and pause at the drone's hover position.

2. Fly up and leftward to about 10–15 feet.

3. **Pause; then fly right in a straight line for about 10 feet.**

4. **Pause; then descend while moving horizontally right.**

   The goal is to end up back at the takeoff hover position.

REMEMBER

As with other shape movements, the triangle will tend to grow and get misshapen as you practice. That's okay. The goal of these exercises is to work on muscle memory with the joysticks. Speed doesn't win the race here — slow down and take your time.

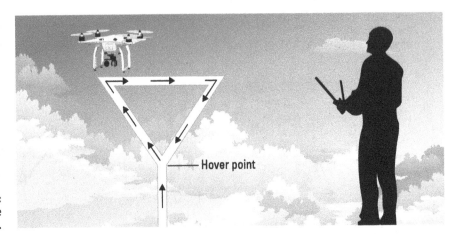

Hover point

FIGURE 3-5:
The triangle
maneuver.

Plus sign

The plus sign maneuver is all about starting and stopping in the right locations, which is (of course) an important skill (see Figure 3-6). Here's how you practice this maneuver:

1. **Find a single object you can hover above, such as a tree; then ascend and get directly above that object (about 15 feet should work).**

   Make sure no other obstacles are in your path at this elevation.

2. **Point the camera angle straight down so that you're looking at the top of the object.**

3. **Fly in a straight line for about 20 feet and then stop.**

4. **Fly backward, stopping over the top of the tree.**

5. **Continue flying backward for 20 feet and stop.**

6. **Fly forward to the tree and stop again.**

7. **Turn 90 degrees (a right angle) and repeat Step 3.**

When you're done, you've drawn a plus sign in the sky!

TIP

The main goal of this exercise is to start and stop at the right places. Try to make sure that you stop directly over the center object point each time and in the same place.

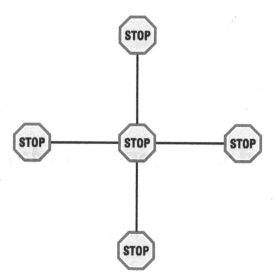

FIGURE 3-6:
The plus sign maneuver.

Stairs

The final basic maneuver you should practice is stairs. In this maneuver, the drone climbs an imaginary set of stairs in the sky and comes back down again (see Figure 3-7). This maneuver helps you practice forward and vertical line movements. Do this:

1. **Fly up to about five feet.**

2. **Fly forward five feet.**

3. **Repeat Steps 1 and 2 four times.**

   Doing so creates a stair pattern.

4. **Turn the drone around and fly back down the stairs, repeating the pattern in reverse.**

   The goal is to end in about the same place you started.

TIP

A sports field is a great place to practice this and other movements. Also, consider picking up some inexpensive traffic cones and placing them on the ground to use as visual marks for the stops. The cones can help you with visual accuracy. You can find practice cones at a sports goods store or Amazon.com.

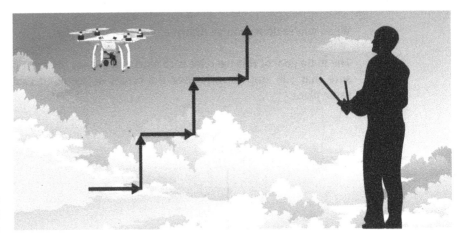

FIGURE 3-7:
The stairs
maneuver.

Pushing Forward with Complex Maneuvers

The basic drone maneuvers that you practice in the previous section help you gain muscle memory with your joysticks. The movements you perform with basic maneuvers are the kinds of maneuvers you'll use regularly as you fly.

As you continue to practice the basic maneuvers, start adding more complex maneuvers to your repertoire. Some of these aren't necessarily difficult, but they require more finesse with your joysticks to get smooth video footage.

REMEMBER

For these maneuvers, you'll record video so that you can take a look at the maneuver afterward. The default video settings are all you need for practice. Be sure to check your drone's documentation for details about transferring video files from your drone to your phone or computer.

All right, get ready to fly!

Spiral

The spiral pattern is just as its name suggests. You'll fly in a spiral as you descend. To do this maneuver, use the RC to fly up to about 350 feet. Turn on your video recording and begin descending as you move the left joystick slightly left or right (called *yaw*), depending on the direction you want to spin (see Figure 3-8).

The key to this maneuver is getting the right descent and yaw speed working together so that you have a pleasing spin at a speed you want. Try avoiding any jerky movements that interfere with the spiral pattern. You can record different speeds and compare them later to see which speed you believe to be most effective.

TIP

As you practice, try some variations:

>> **Reverse the direction.** Create a spiral flying down or flying up. Also, try the same pattern with yaw to the left or right, which spins the drone in different directions.

>> **Try less spin and more horizontal movement.** The video will show more vertical movement (up or down) but less spin.

>> **Try more spin and less horizontal movement.** This technique will give your video more spin but less up or down movement. Remember to experiment at different speeds on the spin.

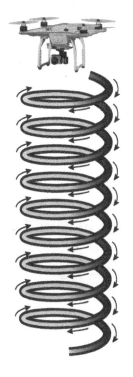

FIGURE 3-8:
The spiral
maneuver.

Orbit

Using the orbit maneuver, you orbit around a structure, such as a water tower or tall building. To orbit around the structure, you use both the left and right joysticks so that you fly sideways in one direction while also turning the drone at the same time. This movement creates a circle pattern, and the idea is to keep the subject in the center of your screen as you fly and shoot the video (see Figure 3-9).

Small movements on the controller make a big difference, so you'll have to experiment to get the correct motion on the two joysticks so that you orbit the structure you choose. Also, be careful with the joysticks so that you don't gain or lose any elevation as you fly.

WARNING

You'll be flying sideways for this maneuver, and many drones do not have side sensors. If yours doesn't, you won't have any obstacle avoidance, so use extreme caution. Make sure the flight path is clear of anything you might run into.

TIP

As you practice, try some variations:

>> **Reverse the direction.** Try flying half the orbit in one direction, and then stop and reverse the direction. This is good joystick practice.

>> **Try to orbit but also gently descend or ascend at the same time.** Again, small movements make a big difference, but the vertical movement can add a new layer to your orbit video.

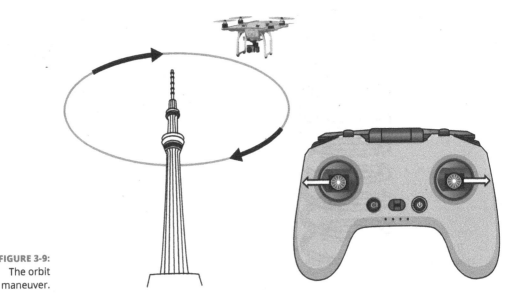

FIGURE 3-9:
The orbit maneuver.

Fly with gimbal rotation

A *gimbal* is a device attached to the drone's camera that manages the camera's movements. Your drone's gimbal can likely rotate up or down using a control on the RC. Often, this is a wheel control on the back of the RC, but RC models vary, so check it out. With this maneuver, the idea is to select a subject below your flight path and, while flying toward it, slowly rotate the gimbal down, keeping the object in the center of the screen as you fly over it (see Figure 3-10). This is great practice for managing the gimbal control while you fly.

As you practice, try some variations:

>> **Try the movement at different speeds.** Try flying slowly and rotating the gimbal slowly. Then, fly the path again with a faster forward speed and faster gimbal speed.

>> **Try the movement in reverse.** Try starting with the gimbal pointing straight down at the subject, and then fly in reverse, slowly raising the gimbal at the same time. Try this move at different speeds as well.

>> **Try the movement while adding altitude or yaw changes.** You can also try this movement while slowly gaining or losing altitude and with yaw changes, which will add rotation to your video at the same time. This movement is more complex because your brain will be going in several different directions, but it's great practice!

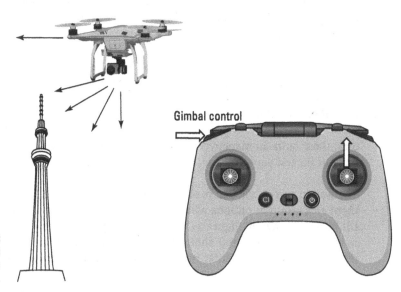

FIGURE 3-10:
Fly with gimbal rotation maneuver.

Gimbal control

Flying with Built-in Maneuvers

Although the built-in maneuvers you can use depends on the drone you own, the odds are good that you can take what are often called *quick shots*. For example, many drones have a built-in orbit maneuver. You simply get your drone in the correct location for the video shoot, turn on the built-in maneuver, and the drone flies the orbit and records the video.

Wait a minute! If the drone will do these kinds of maneuvers automatically, why learn to do them manually? The answer is simple: You should be able to pilot the drone without depending on software. The more you can do yourself, the more flexibility you'll have when you fly.

Still, these built-in maneuvers are great, and they save time, so if you have them on your RC and you need one, then by all means use it!

The way to access built-in maneuvers depends on your RC. Generally, you'll find them as a group of options when you select video shooting. For example, on the DJI RC Pro, shown in Figure 3-11, when I choose Video, I can then choose from a number of preconfigured shots. In Figure 3-11, I have selected Dronie, which means the drone will center on the subject and fly backward and upward away from it. So I choose a subject by selecting it on the RC screen. After I choose the subject, I can also choose how far away from the subject the drone will fly backward. Then I press Start, and the drone records the video, flying backward at an incline while keeping the subject in the center. When the shot is done, the drone automatically returns to the starting position. Easy!

Dronie option   Video selected

FIGURE 3-11:
Built-in
"dronie"
maneuver.

TIP

It's a good idea to experiment with any built-in maneuvers your drone provides. You'll probably find some more useful than others, but you should know what movements are preconfigured and available to you when you're flying and shooting.

Shooting with Active Tracking

Active tracking is a kind of built-in maneuver that your drone may support (see the documentation). With active tracking, you select a moving subject as you're flying, and the drone will *track*, or follow, it.

Why not just track it yourself without this automated feature? Here's the deal: To track a subject, you have to maneuver the drone so that the subject stays in the center, or main focus, of your video (hard to do). At the same time, you have to try to keep a consistent speed because slowing down and speeding up won't look good on video (even harder to do). You have to do all of this while also doing everything else necessary to fly the drone at the same time (hardest). Sure, it can be done, but you'll get better results using the active tracking feature.

Aside from tracking a subject, you can use active tracking to create a video of you doing something, such as driving a car or riding a bike. You wouldn't be able to do this by yourself without the active tracking feature. (If you can ride a bike and fly a drone at the same time, you have some crazy skills!)

TIP

Depending on your RC, you may need to turn on tracking. This is typically a menu option that you can select. For some RCs, it's available by default. All you have to do is select an item on the screen with your finger by swiping over it to create a selection box. Note that on most drones, this feature works only when the drone is in flight.

Many drone manufacturers provide a few different active tracking options. Here are the DJI options, but check your drone documentation for specific details:

>> **Active Track:** This option tracks a moving subject from the back, front, or the side. You choose the option you want when you select Active Track. Remember that if you choose a side option, you won't have any crash protection if your drone doesn't have side sensors. When you active track, the drone will keep the subject at a constant distance and altitude.

>> **Spotlight:** Using this mode, the drone stays in the same position and tracks a subject without physically following it. You can still fly the drone at the same time while the drone keeps its focus on the subject you select. For example, in Figure 3-12, I have the Spotlight feature on a truck.

>> **POI (Point of Interest):** The drone orbits a moving subject, but you can also change the elevation and distance while the tracking is in progress.

Truck is selected

FIGURE 3-12:
The Spotlight feature tracks a subject without physically following it.

Spotlight is selected

Using DJI MasterShots

While I'm on the subject of built-in maneuvers (quick shots) and active tracking, if you use a DJI drone, you probably have one more level of automation you can put to work. DJI has a feature called MasterShots (available on the Mini 3 series and up). This automated feature basically creates a series of different quick shots based on the subject you select. The drone films all the quick shots and then automatically puts them together in one video file. This is a great feature if you have a subject you want to shoot while catching it from every angle and with numerous built-in maneuvers. With MasterShots, the software makes those decisions for you and shoots the best quick shots for your subject.

The DJI has three different kinds of MasterShots:

» **Portrait:** If you select a person on the RC (yes, the software can tell whether or not you've selected a person), the MasterShots will configure a series of angles and movements based on that person's location, distance, and other factors. Basically, the selected person is the star of the show. MasterShots performs a series of movements around that person.

» **Proximity:** If you select a subject that's not a person, such as a monument, tower, or other object, MasterShots will configure a series of shots around that subject, based on distance, the subject's height and size, and so forth. For example, in Figure 3-13, I have selected a water tower for a MasterShots shoot.

Subject selected   MasterShots selected

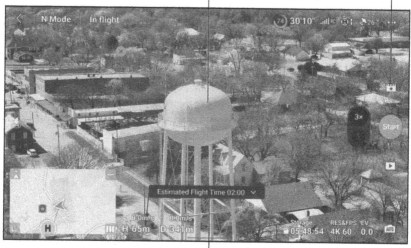

FIGURE 3-13:
MasterShots
proximity
shooting.

Flight time estimation

>> **Landscape:** This MasterShots mode works if you select a larger landscape area, such as a waterfall. The RC will make flight decisions based on the subject's size, distance, and other factors.

When you select a subject for a MasterShot, you can also click the map to see the flight path that the MasterShot will fly (see Figure 3-14). Overall, using Master-Shots is a great way to capture a lot of angles and movements quickly when you want a video of a particular subject.

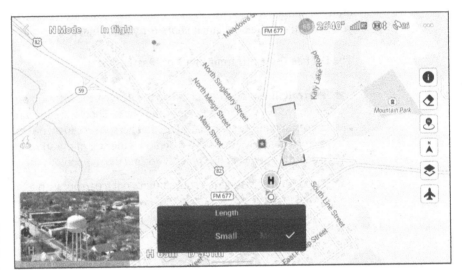

FIGURE 3-14:
A MasterShots map of its flight path.

A WORD ABOUT USING VIDEO FOOTAGE

If you're just getting started with drone piloting, you may be worried about automated video footage. For example, you may think, "I like the dronie shot I took, but I want to fly *toward* a subject, not away from it." Or, "I like the MasterShots footage, but not all of it." Don't worry. You can edit any of this video with a video editor. You can reverse any shots you take so that they are seen "backward" on your final video. You cut sections of video, and trim it, and you can cut, edit, and delete any section of a MasterShot. If all this sounds daunting, you may find that it's easier than you think! Check out Chapter 18 for more information about shooting and editing video.

Growing Your Skills and Confidence

All right, this chapter is loaded with information about flying your drone. If you feel a bit overwhelmed, don't worry; feeling overwhelmed at first is normal! It takes time to build your piloting skills and confidence. As you practice, keep these quick tips in mind:

» **Always perform a pre-flight and post-flight check.** Make sure that you're good to go before you fly. Take a look at Chapter 2 for some helpful pre-flight and post-flight checklists.

» **Always check the weather.** Before you fly, make sure you check the weather forecast and pay attention to the wind. Remember: Wind is not your friend!

» **Focus on a few skills at a time.** Spend a week or so just working on a few foundational skills and maneuvers. Then, add a skill or two and keep practicing them all as you continue adding more. You're working to build muscle memory with the RC joysticks. The more muscle memory you have, the easier piloting will be and the more confidence you'll gain.

» **Don't stress if you have a bad flight day.** By bad flight day, I mean that you didn't perform the maneuvers well or couldn't get the shots you really wanted. Relax. Some days you'll fly better than others. Put the drone away and come back to it the next day. Your piloting performance will be better some days than others.

» **Try to fly every day.** The more you fly, the better you'll be. Weather permitting, try to fly every day as you learn. Even a short 15-minute flight helps grow your skills and confidence. Great musicians and athletes are great because they practice a lot. Great drone pilots are the same.

Chapter **4**

Taking Care of Your Drone and Flying Safely

D rone pilots love to fly. So, when you encounter a chapter about taking care of your drone and general safety practices, it's tempting to skip over it for something more exciting.

However, keeping your drone in good working order and implementing foundational safety practices has everything to do with what kind of flying experience you will enjoy. The good news is, your drone doesn't need a lot of upkeep interaction from you. There are just a few things that you need to regularly check, and they won't take up much of your time, but you do need to read this chapter to see what those regular checks involve.

Also, safe flying is critical for every drone pilot. This chapter also looks at some foundational safety principles that the FAA expects you to know. There are also many additional FAA rules that you can check out in Chapters 6 and 7.

When you first set up your drone and RC, you will likely connect it to your home or office internet-connected Wi-Fi. You'll need that connection for any updates that are necessary. However, it's not uncommon for the drone to forget the Wi-Fi connection after you've used it. You may experience this problem, so it's a good idea to check your controller periodically to verify that it's still connected to Wi-Fi.

Keeping Current with Updates

When you set up your drone and controller for the first time, you likely needed updates (see Chapter 2). However, even after you install those initial updates, you're not done. Your drone manufacturer regularly releases updates that you need to download from the internet and install on your drone and RC. There are two basic types of updates:

>> **Firmware:** *Firmware* is a type of software that is specific to the machine components in a drone and RC. Firmware enables these components to communicate with each other and work together. Without firmware, your drone would just be a bunch of parts that can't interact with each other. Firmware updates are released to improve that communication. Often, the release involves a new software feature or an optimization of an existing one. Also, firmware updates are released to fix known problems or software bugs that are discovered in the previous firmware release. Also, note that some drones require updates on each battery.

>> **Software:** Your drone RC uses software. In this case, the software is sort of like a computer's operating system that lets you see information on the RC screen and interact with it. As with firmware, software updates are released to roll out new features or fix problems with existing software.

Because both firmware and software updates may resolve potential problems, you'll want to keep your drone and RC updated. To do so, you need to get your drone connected to Wi-Fi. The RC typically tells you when new updates are available, but you also may need to check one of the menus. See your drone documentation for details.

You'll be glad to know that firmware and software updates usually take only a few minutes — and those updates are rarely an emergency. If you're getting ready to fly and discover that updates are needed, you can wait until after the flight to download and install them. Just remember to get those updates taken care of soon after they are released. Keeping things up to date will ensure that your drone and RC remain in good working order.

Inspecting and Changing Propellers

Your drone's propellers have everything to do with your drone's ability to fly and fly correctly. Propellers are one component of the drone that you should check constantly — before and after every flight. After all, they're just plastic, and they can easily get damaged.

Inspecting propellers

Before and after you fly, check the propellers for

- » **Nicks:** If the drone propellers hit something on takeoff or landing, they get little nicks or cuts in the edges. These nicks can impact the drone's flight and your ability to control the drone. The propellers in Figure 4-1 have nicks in the edges and need to be replaced. Nicks like these can cause the drone to experience excessive vibration, potentially making a flight unstable.

- » **Scratches:** The propellers can also pick up scratches from hitting items on takeoff or landing. If the scratch is small, it's probably fine, but a larger scratch can mean that the impact that caused it may have altered the shape of the propeller. In that case, replace the propeller.

  Misshapeness: If you hit something on takeoff, landing, or even in flight, the impact can cause the propeller to get slightly bent or misshapened in some way. A misshapened propeller can absolutely cause in-flight problems. The best method of inspection is to slide the propeller between your index finger and thumb. You can often feel bumps or changes of angle that you may not notice with a simple visual inspection. If something feels wrong, change the propeller before you fly again.

- » **Old age:** Manufacturers suggest that you change the propellers after so many hours of use, just due to wear and tear. These recommendations vary based on your drone, so check the manufacturer's recommendations for your drone model.

Nicks like these can destabilize a flight

FIGURE 4-1:
Damaged drone
propellers.

boyloso/Adobe Stock Photos

TIP

If you're flying your drone and you notice that the image on the RC seems jerky or out of alignment, that's a good sign that you need to change the propellers. Another sign is when the drone's batteries suddenly don't last as long as they used to, even in good flying conditions. Bad propellers make the drone work harder to maintain flight direction and speed.

Changing propellers

To change the drone's propellers, you need to follow the manufacturer's instructions for your particular drone model. Different drone models use propellers that are attached in different ways, so check out the instructions and don't try to figure it out yourself!

TIP

If your drone uses a screwdriver to change the propellers, one that fits the propellers' screws was likely provided by the manufacturer. However, a common online complaint is that these screwdrivers are not great. You might consider spending $10 or so and buying a better screwdriver. Make sure to buy the exact size needed for the propeller screws, however, or you can strip the screws as you're trying to get them on and off.

WARNING

If you own a popular drone model, you may be able to find propellers that fit your drone but that are manufactured by a company other than the drone's manufacturer. These are known as *aftermarket* propellers and are often less expensive. However, don't use them! They are not identical to the manufacturer's propellers and may not perform as well. Also, using aftermarket propellers can void your drone's warranty, depending on the manufacturer's rules. I strongly recommend that you stick with the propellers made by the manufacturer specifically for your drone!

Cleaning Your Drone and Camera

No matter how hard you try, your drone will collect dust and dirt on takeoff, landing, and in flight. To keep your drone and camera in good working order, spend a couple of minutes cleaning the drone after each flight. Fortunately, cleanup is quick and easy.

Cleaning the drone body and propellers

Tidying up your drone simply requires you to gently wipe down the body and propellers with a microfiber cloth. A microfiber cloth is best because it's made from very small fibers, which do a much better job of trapping dirt when you wipe down the drone. Also, microfiber cloths don't leave behind fuzz or lint as other kinds of cloths do. Microfiber cloths are cheap and washable, so you can just pick up a package of them and put them to work.

If the drone body picks up some mud or other debris that doesn't easily wipe off with a microfiber cloth, you can use an isopropyl alcohol pad to gently scrub the dirty area. Cleaning pads for eyeglasses also work well. Sure, you can use a slightly damp cloth, but it's best to avoid using water.

Finally, you can also use a bit of compressed air around the seams or other areas that you can't wipe down. But use caution — you don't want to use a lot of compressed air with high pressure on the drone body or other components, and never use compressed air around the camera.

Some drone motors need lubricants from time to time. Check the manufacturer's instructions for details. However, if the manufacturer doesn't say to add lubricants, then don't do anything!

Cleaning the drone camera and gimbal

Your drone's camera uses a gimbal, which is basically a stabilizer, as explained in Chapter 1. The gimbal allows your camera to move as the drone flies but keeps the horizon straight and gives you that smooth video footage you want. Without the gimbal, the camera's footage would be a jerky, crooked mess.

With that said, the gimbal is a delicate piece of machinery, so when you clean the drone camera and gimbal, you want to be gentle. The good news is that camera and gimbal cleaning is not difficult. Generally, you can just wipe the camera lens with a microfiber cloth and wipe the exposed gimbal parts in the same way.

To do a better job with cleaning, consider picking up a bulb air blower and a lens-cleaning pen, as shown in Figure 4-2. The bulb air blower gives you the ability to use gentle puffs of air to blow out any dust around the gimbal. The lens-cleaning pen has a brush but also a nonliquid cleaning end that you use to wipe down the front of the lens. The lens pen is great for removing smudges and fingerprints you might get on the lens.

You can get a blub air blower, lens pen, and other little cleaning items in a kit. Just search the internet for "camera cleaning kit" to find some inexpensive options. These items are good to have on hand.

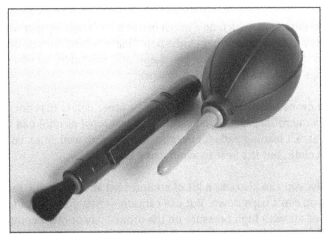

FIGURE 4-2:
Bulb air blower
and lens pen.

Dkoa Galery/Shutterstock

Calibrating the Drone

Your drone has a built-in electric device called an *Inertial Measurement Unit (IMU)*, which is a type of sensor system that helps the drone use an internal compass and other tools to enable the drone to fly correctly. Basically, the IMU ensures that the drone maintains the correct orientation in a given environment. Without the IMU, the drone wouldn't be able to understand what's up, down, and so forth.

Calibration is the process of correcting errors and problems that occur in the IMU system. Without calibration, and over time, the IMU system can experience problems, which may degrade overall flight performance.

REMEMBER

Luckily, calibrating the drone is easy, and you don't have to know the technical details. You calibrate the drone using the RC. You may see a message telling you that the drone needs calibration, or you can typically access the option in the menu system. All you need to do is locate the calibration option and then follow the instructions. If you're unsure, see your drone manufacturer's recommendations concerning calibration and how often you should calibrate the drone. You may also have different calibration options for the compass and gimbal.

For example, in Figure 4-3, I've accessed the gimbal calibration option on the RC. The instructions tell me to place the drone on a flat surface and click Auto to start the automatic calibration. That's it. When the calibration is complete, you're good to go!

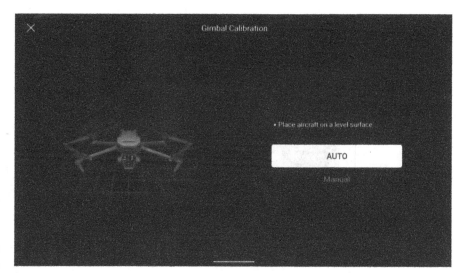

FIGURE 4-3:
Calibration.

Taking Care of LiPo Batteries

Your drone likely uses LiPo batteries to fly. LiPo, which stands for *lithium-ion polymer*, is a type of rechargeable battery that provides more power than a standard lithium battery. Much larger versions of LiPo batteries are also used in electric vehicles. Your drone could use lithium-ion (Li-ion) batteries instead, which is another type of rechargeable battery.

Here's the skinny: LiPo positive and negative ions disperse energy. They are very effective for use in power drones and other remote-controlled devices for a few reasons. First, they're relatively lightweight, which is mission critical for drone flights. Second, they can be manufactured in many different sizes and shapes. For example, Figure 4-4 shows the batteries for the DJI Mavic series. But your drone's batteries may look completely different from these. The capability to manufacture different sizes and shapes of LiPo batteries is critical for the drone industry. The third reason LiPo batteries are so effective is that they have high discharge rates, which means they can give your drone the power it needs to fly and maneuver.

Happily, LiPo technology is safe, and your drone batteries will likely last you a long time.

REMEMBER

The bad news is that if something goes wrong with one of your LiPo batteries, there is a fire danger. Generally, if a LiPo battery is becoming unstable, it will swell. For example, Figure 4-5 shows a normal battery (left) and one that has started swelling on the sides and bottom (right). If you see any swelling on a LiPo battery, get rid of it and buy a new one (to safely get rid of LiPo battery, here are instructions: `https://www.droneblog.com/dispose-lipo-drone-batteries/`). Sure, they're not cheap, but the expense of a new battery is much better than a fire.

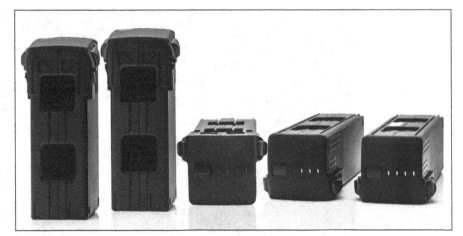

FIGURE 4-4:
LiPo batteries can
be manufactured
in many shapes
and sizes.

Artinun/Adobe Stock Photos

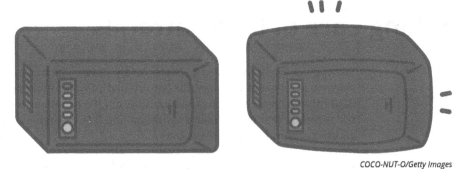

FIGURE 4-5:
Discard any LiPo
battery that is
swelling.

COCO-NUT-O/Getty Images

Aside from watching for swelling or other visible defects, there are just a few quick things you should know about taking care of LiPo batteries:

>> **Use the manufacturer's charger.** You can buy various aftermarket battery chargers specific to your drone's battery. These may be fine — or not; you really don't know. By far the safest approach is to stick with the manufacturer's charger.

>> **Don't charge batteries in hot environments or in direct sunlight.** Drone batteries get warm as they charge — they don't need any environmental heat added to the mix. Charge them in cool environments and out of direct sunlight.

>> **Don't charge immediately after flying:** Drone batteries get hot as they disperse energy. After flying, let the battery cool down for a half-hour or so before charging.

STORING YOUR DRONE

When your drone and drone batteries are not in use, store them in a cool, dry place away from sunlight. A closet is, of course, a good location. Also, it's a good idea to get a hardshell backpack or case for your drone. These products protect the drone should the case or backpack get dropped, knocked around, or stored under other objects. That way, when your teenager decides to throw a bag of cat food on top of your drone stored in the closet, you'll know that the hardshell case is keeping it protected!

» **If you're not going to use your drone for a while, store the batteries with a 60 percent charge.** This doesn't have to be exact, but strive for 60 percent to maintain good battery life. If you don't plan on using the batteries for several months, periodically recharge them to about 60 percent. LiPo batteries need to maintain some charge, or they may quit working.

Troubleshooting Common Problems

Like all electronics, your drone isn't perfect and can experience problems from time to time. You've already explored most of the most common drone problems in this chapter: firmware updates, propeller issues, calibration errors, and battery problems. Regular maintenance of your drone, RC, and batteries can prevent most of issues.

The following sections point out a few more common challenges.

Environmental effects

Your drone can withstand hot and cold environments — within reason. Drones generate a lot of heat when they fly, so if you're out on a hot day in direct sunlight, you may experience overheating and shorter battery life. The same is true in very cold environments. The battery will discharge much faster when the temperature is cold, and mist in the air can even freeze on the drone, which adds weight and can affect the propeller performance. So use caution in hot and cold environments, and find out more about various weather issues in Chapter 14.

Camera issues

If you're having problems with the camera, there are a few things to check. First, make sure the firmware is updated. Second, calibrate the drone (see the "Calibrating the Drone" section, earlier in this chapter). Third, check the camera settings. You may have accidentally switched from an Auto mode to a Manual mode. Make sure the camera settings are what you want for your needs and skill level. Also, see Chapters 17 and 18 for more information about drone photography and videography.

Connectivity glitches

When you fly, your drone depends on GPS signals and radio signals from your controller. You can lose either signal for a variety of reasons, often beyond your control. In these cases, you want your drone to be able to come home on its own, which is why RTH (Return to Home) is such an important feature. If you continue to have signal problems, this is probably a sign of something internal that needs to be repaired.

Getting help

You may experience problems that are beyond your ability to troubleshoot and fix. In this case, you'll need help from the manufacturer. Check out the drone manufacturer's website for support options.

TIP

Not all drone companies provide the same level of support. Frankly, some are better than others. If you're in the market for a new drone, you should read reviews of the support and warranty options the manufacturer provides before you make a purchase. Check out Chapter 1 for more information.

Exploring Foundational Safety Practices

Along with taking care of your drone and troubleshooting problems, following foundational safety practices can also help you keep your drone safe along with people and property.

The FAA puts forth numerous drone regulations, and you can check out the details of those in Chapters 6 and 7. For now, it's important to review basic safety principles you should employ at all times. The FAA expects you to fly with these safety principles in mind, so be sure to commit them to memory.

Maintain visual line of sight

The FAA requires drone pilots to maintain visual line of sight (VLOS), also mentioned in Chapter 2. VLOS means that you can see your drone as you fly. The FAA allows a "visual observer" to stand with you as you fly and help you maintain VLOS as you use the controller. Although some drones, such as the DJI Mini series, are small and difficult to see with much elevation, the FAA maintains this rule for safety. If you can't see the drone, the odds go up that you'll have some kind of accident. So, always keep your drone in your VLOS.

Fly in authorized airspace

The National Airspace System (NAS) has numerous rules about piloting a drone in various classes of airspace. The FAA expects you to know this information and use a B4UFLY app (see Chapter 2) to check flight restrictions. You can learn more about NAS in Chapter 9.

Do not fly in a dangerous manner

Generally speaking, a dangerous manner of flying involves flying too low, too high, or too fast. Many drones do not have sensor protection in Sport or Speed mode, so it's easy to run into something or someone using these modes. Although it's fun, play it safe and do not fly too fast, too high (out of the airspace requirements), or too close to buildings, trees, or people.

Stay away from emergency scenes and personnel

A traffic accident, a fire, a hostage situation — whatever the case may be, if there are law enforcement and emergency personnel on a scene, stay away. It can be tempting to fly close to such scenes and record video, but this action violates FAA rules. Avoid these situations.

WARNING

Emergency situations may include other air traffic, such as helicopters. Even if the emergency is happening in an airspace in which you're allowed to fly, the emergency situation takes precedent. You don't want to cause a helicopter crash with your drone.

Do not fly under the influence of drugs and alcohol

Drugs and alcohol use can greatly affect your ability to pilot a drone. Even some prescription medications can pose hazards. Never fly under the influence. You can find out more information about drugs and alcohol use in Chapter 7.

Don't talk to people when you fly

This safety rule may seem peculiar, but you shouldn't talk to other people when you fly. Talking to someone creates a distraction, and you want all your mental focus on your drone and RC. If you're flying in a place with other people around, such as a park, keep in mind that people are naturally curious. A bystander may approach and ask you about the drone, what model you fly, and so forth. It's a good idea to simply say, "I'm not allowed to talk to people when I fly. I can chat with you after I land." This may seem a bit over the top, but think of it this way: Do you really want to have an accident because you were talking to a stranger?

Keep it safe — save the conversations until after you land.

2

Following FAA Regulations

IN THIS PART . . .

Explore Part 107 Certification and TRUST certificates.

Get the details of FAA flight rules.

Understand additional FAA regulations.

Fly your drone safely at night.

Chapter **5**

Exploring the TRUST Certificate and Part 107 Certification

In the world of drone piloting, you often begin experimenting with an inexpensive drone just for fun. Suddenly, you realize you enjoy the hobby, and a better drone would be . . . well, better! So you upgrade, and then you want to upgrade again.

Before long, you start to have some nice photos and videos, and someone says, "Hey, do you think you could shoot a birthday party for me with your drone?" or "Could you take a drone photo of our business location so we can put it on our website?"

This series of events is what often leads people into drone piloting as a career. In fact, this series of events is what may have led to you purchase this book!

Now for the question: Do you need Federal Aviation Administration (FAA) certification? Is taking the FAA exam and getting certified to fly your drone necessary, or worth it? Only you can answer the second question, but in the end, the answer is likely to be yes.

In this chapter, you explore the TRUST certificate and Part 107 certification, including categories of information on the exam, tips for studying, and how to register for the test. As you look at what the exam comprises, you may realize that many other chapters in this book explore content found on the Part 107 exam. That's because the FAA expects drone pilots to know all this stuff!

Clearing the Air about TRUST Certificates and the Part 107 Exam

First things first. People experience a lot of confusion about TRUST certificates, the Part 107 exam, and whether you need the Part 107 credentials to fly in various situations. The information you find on the internet may compound this confusion. Keep in mind that a lot of people have opinions on the matter and will post those opinions as fact when, in fact, they're not!

I mention TRUST certificates and the Part 107 exam here and there in this book, but in this chapter I take a moment to clarify the difference between the two. Also, the issue of drone weight and FAA registration gets mixed up in this topic as well. The good news: It's really not that complicated!

Drone weight

Here's the rule: You can fly a drone without FAA registration if the drone weighs no more than 250 grams, or 0.55 pounds, as shown in Figure 5-1. When you're shopping for a new drone, you'll even see the weight issue used in advertising to make the weight seem like more of a hassle than it really is. DJI, Autel, Exo, and many other drone companies all sell drones that fall just under this weight restriction, so you don't have to register those drones with the FAA. It's important to note that 250 grams is the maximum weight. If you upgrade the batteries or propellers, those upgrades may add more weight, pushing your drone over the 250-gram threshold.

However, people often think that because they have a drone that weighs under 250 grams, they can do whatever they want. Not true! You still can't use this drone for any commercial purpose without a Part 107 license. Weight and commercial use are two separate issues.

Register with the FAA Drone Zone

250 Grams

Registration not required

FIGURE 5-1:
Drones weighing over 250 grams must be registered with the FAA.

Fortunately, if you buy a drone that weighs more than 250 grams, it's really not a big deal. You must register it with the FAA and pay a $5 registration fee. Registration takes only a few minutes. Just head over to the FAADroneZone website at https://faadronezone-access.faa.gov/, create an account, and then follow the prompts to register your drone. After you register, you receive a registration number for the drone that you need to affix to it using a permanent marker, an adhesive label, or another permanent method. This number will be your drone's registration number for life. That's all there is to it.

REMEMBER

Be aware that registering your drone registration and licensing it to fly for commercial purposes are two different things! The question is, how do you plan to use your drone? That's where licensing comes into play, so read on.

Recreational flyers and TRUST certificates

The FAA has two foundational categories of drone pilots: recreational pilots and certified remote pilots. A *recreational* pilot is someone who flies for fun. You simply want to have fun flying your drone and taking photos and videos for your personal use. You can post your photos and videos to your social media accounts (provided that those accounts aren't monetized) so that you can share them with your friends and family.

But when it comes to what you can do with your drone, that's it.

You can't do anything else with your drone, for all practical purposes. You can fly it for personal enjoyment and nothing else.

Even if you're strictly a recreational flyer, however, the FAA still expects you to know and follow some foundational safety rules and procedures. I list those in Chapter 2, so be sure to check them out.

Also, even as a recreational flyer, the FAA requires you to take the Recreational UAS Safety Test (TRUST). The TRUST test is free, and it tests your knowledge of basic drone safety information. Drone piloting is popular, and the more drones there are in the air, the more likely it is for accidents to occur. The FAA wants drone pilots to fly safely, so the TRUST test checks your knowledge of foundational safety principles. Law enforcement or FAA personnel may ask to see your TRUST certificate anytime you fly.

So what's on the TRUST test, exactly? This test asks you basic recreational flight safety questions, which you can review at https://www.faa.gov/uas/recreational_flyers; also see Chapter 2. Luckily, when you want to take the test, you'll find a 30-minute course, and then you can take the test based on the information you just learned. You can take the test as many times as you want until you pass. To get started, just go to https://www.faa.gov/uas/getting_started. There, you can find several courses and test providers. They all do the same thing, so just pick one. And remember, the courses and test are all free.

TIP

After you pass the test, you should keep your TRUST certificate with you when you fly. But remember this: The FAA or the testing website doesn't keep track of your TRUST certificate. If you lose it, you'll need to take the TRUST test again to get another certificate.

Certified remote pilots

After you move beyond the recreational flyer level, you enter the realm of certified remote pilots, and that's where the Part 107 exam comes into play.

The FAA expects certified remote pilots to know and understand the foundational recreational flyer rules plus a few more. Check those out in Chapter 2. Note that you may get your TRUST certificate and then move on to the Part 107 exam so that you have both certifications, though Part 107–certified pilots are not required to have a TRUST certificate.

REMEMBER

Also, after you become a certified remote pilot, the drone registration rules change. You must register any drone you use with the FAA if the drone is used for commercial purposes, even if the drone weighs under 250 grams. That's not a big deal, and registration cost remains $5 per drone whether you're a recreational flyer or certified remote pilot.

With that said, the rest of this chapter focuses on the certified remote pilot and the Part 107 exam. Read on!

Determining Whether You Need a Part 107 Certification

To find out whether you need a Part 107 certification, you need to think about how you use your drone and how you may want to use it in the future.

Here's the rule: If you do anything with your drone that is not purely recreation for your personal enjoyment, you need a Part 107 certification. Often, the confusion on this point revolves around money. Online, people often proclaim, "As long as you don't make money with your drone, you don't need the Part 107."

That's not true.

Money has nothing to do with the FAA's requirements. It's all about the commercial intent of your piloting. The easiest way to understand this is to look at some examples. The following sections do just that.

Beach photos

Your family takes an annual trip to the beach. You want to fly your drone to take some photos of your kids playing, as in Figure 5-2. You'll later share those photos on your social media, but your social media isn't monetized. You're taking these photos for your memories and to share them with Aunt Wanda who is always bugging you for more pics of your kids. Do you need a Part 107 license?

No. This photo shoot is for recreational purposes and the photos are for personal use.

Family reunion

You've been asked to take some drone photos of your family reunion to be held at the park. Afterward, you'll give copies of your best shots to your family members to enjoy and keep. Some family members may share the photos on their social media accounts. You're quite certain that Uncle Randall will post them on his Facebook page because he posts every single thing he does there. Do you need a Part 107 license?

Tomsickova/Adobe Stock Photos

No. This photo shoot is for recreational purposes and the photos are for personal use.

Small business photos

You have a friend who owns a small gardening center. Your friend asks you to take some aerial photos of the garden center (which might include shots like the one in Figure 5-3) so that she can post them on her website and social media. She's a friend, and no money will exchange hands for the photos. Do you need a Part 107 license?

Yes. This photo shoot is for commercial purposes. The issue of payment is irrelevant. The photos will be used for commercial purposes, so you need a Part 107 license.

FIGURE 5-3:
Business photos
require a Part
107 license.

Nonprofit photos

A church in your town is hosting a city-wide Easter-egg hunt at the park. You've been asked to take some photos of the hunt for the church's web page and the Chamber of Commerce. You won't be paid for these photos, and you'll keep a reasonable flight distance from the kiddos to make sure everything is safe. Do you need a Part 107 license?

Yes. This photo shoot is for commercial purposes. Nonprofit organizations fall under the category of "commercial." The issue of payment is irrelevant. The photos will be used for commercial purposes, so you need a Part 107 license.

Friends building a house

You have some friends who just bought ten acres of undeveloped land. Your friends want to build a house on the property, but they're unsure of the best location for the house. They've asked you to fly over the property, taking photos and videos so that they can get an aerial look. They'll pay you $100 for this quick shoot. Do you need a Part 107 license?

Yes. This photo shoot is for commercial purposes because you're being compensated. You need a Part 107 license. Although payment alone isn't a determining factor in defining "commercial," if you are paid, you need the Part 107 license.

Insta-famous mountains

You love mountain hiking, and last year, you started an Instagram page about hiking various mountain ranges. Your Instagram has been successful, and you've managed to monetize it. You want to take some aerial photos of a hike that is coming up and use those on your Instagram page. Do you need a Part 107 license?

Yes. This photo shoot is for commercial purposes because your Instagram is monetized. You need a Part 107 license. Also, say you took some great vacation photos before you had your Part 107 license, but after getting your license, you want to use them on your monetized social media. Can you? No, because you took the photos before you were certified.

TIP

From the examples you've read, you probably get the picture, but Table 5-1 helps to clarify as well. You can't do much of anything outside of personal, recreational flights without a Part 107 certification. This is why most people who enjoy flying drones eventually start looking into getting certified.

TABLE 5-1   Drone Photo/Video Activities That Require Part 107 Certification

Activity	Needs certification
Yours or anyone's personal use only (including noncommercial social media)	No
Paid by anyone	Yes
You're not paid, but you're flying for a business of some kind	Yes
You're not paid, but you're shooting for a nonprofit, which makes it commercial	Yes

Checking out the Part 107 Exam

For the record, the Part 107 exam is actually called the *Unmanned Aircraft General – Small (UAG)*. This is the name of the test you'll actually register for and take. However, everyone just calls it the Part 107 exam, named after the regulations for flying a drone in the United States (14 CFR Part 107). In other words, when you fly, you fly under the Part 107 rules.

You must meet a few basic requirements before you can take the Part 107 exam:

>> **Age:** You must be at least 16 years old to take the Part 107 exam.

>> **Language:** You must be able to read and speak English.

>> **Capability:** You must be mentally and physically capable of safely flying a drone. In other words, you have to be able to make flight decisions and you must be able to control the drone's RC with your hands.

Also, note that the test is 60 multiple-choice questions, and you have two hours to complete it.

Now, the big question is, what's on this exam, exactly? The following sections outline what kinds of content you'll find on the exam, but know up front that this information can a bit overwhelming.

When I first started reading about the Part 107 certification, I quickly thought, "This is way too much trouble!" You may well feel that way, too, but try to push through those feelings. You can learn the content and pass the exam — with some time and effort! In fact, you're learning a lot of the content you'll see on the exam because you're reading this book. So don't let the following sections overwhelm you! You can tackle the exam piece by piece.

Understanding airspace

In the United States, the National Airspace System (NAS) is used to help drone pilots, helicopter pilots, commercial airline pilots, and all other pilots know where to fly. You can think of the NAS as a network of controlled and uncontrolled airspaces. Different rules govern different airspaces for all aircrafts, drones included.

Even if you're a recreational flyer, the FAA expects you to know the basics of airspace and the rules about flying no higher than 400 feet. For the Part 107 exam, you'll need to know much more about the airspace and the restrictions.

For example, here's a sample airspace question: Referring to Figure 5-4, what is the floor of the Class B airspace at Dallas Executive Airport (RBD)?

The answer: Class B airspace at RBD is 3,000–11,000 feet. On the chart, you see the notation that says 110/30, which means the airspace ceiling is 11,000 feet and floor is 3,000 feet. So, the answer is 3,000 feet.

Here's another example: Referring to Figure 5-5, are there any safety concerns with flying a UAS in this location?

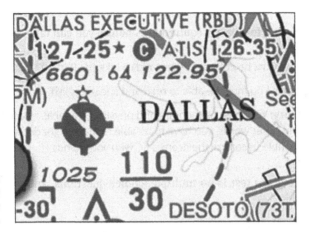

FIGURE 5-4:
Class B
airspace
at RBD.

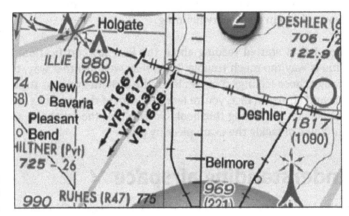

FIGURE 5-5:
Airspace
concerns.

Answer: Yes. In the figure, you see the Victor Route (VR) notations, such as "VR1667." These notations mean that these are military training routes, which could mean traffic below 400 feet. So, this area does pose a danger with flying a UAS.

Now, if these examples sound absolutely horrifying, don't worry. It's not as difficult as you might think after you understand airspace notations on charts. You can learn more about airspace in Chapter 9 and more about sectional charts in Chapters 11 and 12

Checking out flight regulations

There are numerous questions on the Part 107 exam about various flight regulations. You can find out about flight regulations in Chapters 6 and 7, but here's a quick sampling of some types of flight regulation questions you may encounter on the exam:

Question: You experience an accident with a small drone that scratches the arm of a crew member. Do you need to report this accident to the FAA?

Answer: No. There are rules about what kinds of accidents need to be reported. You can find out more about reporting accidents to the FAA in Chapter 7.

Question: What is required to operate an unmanned aircraft (UA) thirty minutes after civil twilight?

Answer: Anti-collision lights. There are rules about operating your drone during daybreak and twilight hours. See Chapter 6 for details.

Question: You've been asked to fly your UA for an inspection job in Class C airspace. How can you fly in this airspace?

Answer: You must receive Air Traffic Control (ATC) authorization from the ATC facility that has jurisdiction over the Class C airspace in question. See Chapter 13 for information.

Dealing with weather

In order to pass the Part 107 exam, you'll need to become at least an amateur weather reporter. You don't need to know everything, but you'll need to know how different kinds of weather conditions affect the flight performance and safety of a drone. You can read about these issues in Chapter 14.

Here are a few sample questions:

Question: What are the characteristics of stable air?

Answer: Poor visibility and steady precipitation. Although "stable" sounds like good air, it actually has rain and poor visibility.

Question: What effect does high-density altitude have on a UA?

Answer: Decreased propeller efficiency. High-density altitude means there are fewer molecules in the air. The drone's propellers use air molecules for lift, so when there are fewer molecules, the drone has to work harder to fly.

Question: You read a METAR that reads 121853Z 18006KT 1/2SM FG R04/2200 OVC005 20/18 A3006. What is the wind direction and velocity?

Answer: The direction and velocity are 180 degrees true at 6 knots. A METAR is a Meteorological Aerodrome Report, which is a kind of weather report format that pilots use. The Part 107 exam will ask you to decipher reports like this, and the good news is that they're not too difficult after you understand the basic parts. The 18006KT part means 180 degrees true at 6 knots. (180 degrees true means the direction relative to true north).

Question: METAR KBOI 121854Z 13004KT 30SM SCT150 17/6 A3015 METAR KLAX 121852Z 25004KT 6SM BR SCT007 SCT250 16/15 A299

According to this METAR, what is the visibility at Los Angeles International Airport (KLAX)?

Answer: Six miles and misty conditions. On the METAR report, you see a notation with some characters after "KLAX" that says "6SM BR." This notation means that visibility is 6 statute miles (a statute mile is one mile over land, or 5,280 feet), and BR means mist (and who knows why BR means mist!).

Again, these questions aren't too hard after you know what the notations mean! The good news is that there aren't a lot of notations to learn, and the testing supplement you'll get when you take the test has a legend that will help you break them down Check out more information about reading weather reports in Chapter 14.

Considering loading and performance

You'll find a few questions on the exam about loading and performance on a drone. These questions have to do with the center of gravity and how different conditions affect performance.

The good news that is you'll probably only see a few questions on this topic, and they're not too hard after you understand the principles. Sometimes you'll even see questions about uncrewed ("unmanned" on the test) airplanes instead of drones. See Chapter 16 for details. The following are couple of sample questions:

Question: When should a remote pilot consider the load factor placed on an unmanned airplane's wings?

Answer: Anytime maneuvers call for anything other than a straight and level flight. By common sense, anytime a flight isn't straight and level, that airflow puts more load on the plane's wings.

Question: Referring to the chart in Figure 5-6, a UA weighs 19 pounds. At a bank angle of 45 degrees, what load will be placed on the UA while maintaining altitude?

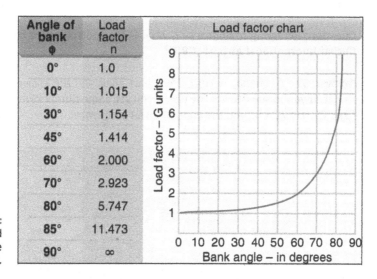

Angle of bank φ	Load factor n
0°	1.0
10°	1.015
30°	1.154
45°	1.414
60°	2.000
70°	2.923
80°	5.747
85°	11.473
90°	∞

FIGURE 5-6: Loading and Performance chart.

Answer: 27 pounds. You'll probably see this chart on the Part 107 exam, and the good news is that the chart basically gives you the answer. A banking aircraft adds load, or weight, to the craft from the pressure of the bank. If you look at the chart, you see that 45 degrees corresponds to a 1.414 load factor. To get the answer, multiply the drone's weight by 1.414, which is 27 rounded up.

Understanding airport operations

This section of the 107 exam is a hodgepodge of content concerning airports, flying around airports, and communications with airports. You may think, "I just want to fly my drone, not deal with an airport!" However, you could end up with a drone piloting job that requires you to fly near airports for a variety of reasons. In this case, you would need to know about airport operations. See Chapter 13 for details. In the meantime, here are a couple of sample questions:

Question: While monitoring airport traffic, you hear an aircraft announce that it is midleft downwind to runway 14. Where is the aircraft relative to the runway?

Answer: The aircraft is in the east. To answer this question, you have to understand that runway numbers are based on compass headings and aircrafts approach runways in the same pattern. When you have the pattern down, these questions are easy to answer because the pattern applies to all air traffic preparing for landing. Check out Chapter 13 for more details and some illustrations.

Question: Referring to Figure 5-7, what is the control-tower frequency at Corpus Christi International Airport (CRP)?

Answer: The frequency is 119.4. The chart gives you the answer; you just have to find it. If you locate CRP on the chart, you see some numbers under the name. The first, CT-119.4, tells you the control tower frequency.

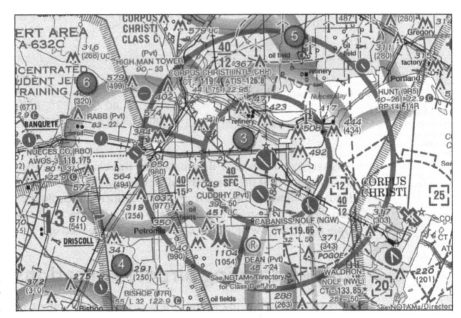

FIGURE 5-7:
Control Tower
frequency at CRP.

Understanding Crew Resource Management

You may think of flying a drone as a solitary activity, but in many commercial cases, more than one person may be involved. The FAA calls these *crew members*, and it's an important concept to understand for real-world drone operations as well as the Part 107 exam.

Crew Resource Management (CRM) refers to managing the resources available to a remote pilot in command. This concept includes crew members but can be used to mean anything that helps you safely complete a drone mission, such as additional software.

A crew, effectively, can be any number of people who help you complete a drone mission. For example, in tricky or difficult flight situations, you could have additional crew members who serve as visual observers, weather reporters, and loading analysts as well as perform other tasks that can be happening while you are flying the drone.

The FAA notes a few different concepts concerning CRM. These are straightforward and fall under the category of best practices when you fly. Consider these points:

>> **Plans and checklists.** The aviation community is a big fan of plans and checklists for a simple reason. Often, aviation tasks are repetitive and become mundane, which makes it easy to drop or forget those tasks. Accidents can occur as a result. So an important part of CRM is making plans and using checklists when you prepare for a drone mission. If other crew members are helping you, plans and lists get everyone on the same page and eliminate any confusing aspects of the mission or its purpose. It makes sense to use a plan, make a checklist, get these items on paper, and review them together.

>> **Communications.** If you're flying a drone mission and you have other crew members, how will you all communicate and to what extent? Having too much communication creates confusion, but having too little fails to fulfill the goal of utilizing crew members. If someone is monitoring airport traffic, plan how that person will give you an update. If someone else is a visual observer who's not standing next to you as the pilot, determine how you will talk to each other. Make a communication plan and organize that plan around the specifics of the particular mission you're flying. Different missions will have different communication needs, so you don't have a one-size-fits-all approach. Tailor communications to the mission.

>> **Task management.** If you have a crew, decide on who is doing what, and be specific. Establish what a visual observer is supposed to be observing. Know what an airport traffic observer is paying attention to, exactly. Each mission has different tasks, but in complex missions, it's important for each crew member to understand their job. That way, one person isn't trying to do too much during the flight whereas another doesn't have enough to do. Make a task-management plan and review it carefully with anyone who is a part of your crew.

>> **Situational awareness.** A big issue with CRM is always situational awareness, and in the end, your plans, communication, and task management all revolve around the mission. You have to think about the situation you're flying in as well as the possible hazards. From weather and air traffic to obstacles, situational awareness means that you are aware of the situation you're flying into. Think carefully about safety issues and how CRM can help you have a safe and productive flight.

Studying for the Part 107 Exam

In the world of taking tests and exams, there is a difference in knowing the content or what I like to call "buzzing the test." In the former, you study the content and then practice with sample exam questions to help you get ready. In the second, you spend all your time memorizing practice questions in hopes of getting a passing score. You don't really know anything but managed to memorize just enough content to pass.

The fact that you're interested enough in drone piloting to be reading this book suggests that you want to actually know the information relating to the Part 107. You want to be a drone pilot who makes good decisions, flies safely, and can adapt to many different piloting situations you may face. In other words, you want to actually know the content the FAA expects you to know.

When it comes to studying for the Part 107 exam, I have some recommendations. First, keep reading this book. Take your time with areas of content that are difficult or confusing, especially airspace and sectional charts (Chapters 9 through 12). That content is often difficult for people, so just keep working on it.

TIP

After you've studied this book carefully, taking a course that helps you prepare for the exam can be a great idea. A course will review the same content that you read in this book and provide quizzes and full-length practice exams. Many courses even provide a guarantee that you'll pass the exam. You have numerous courses to choose from, but I can recommend two that I'm familiar with: Drone Pilot Ground School (www.dronepilotgroundschool.com) and pilotinstitute.com (www.pilotinstitute.com). Both are good courses that can help you get ready for the test.

WARNING

You can also access many YouTube videos for free. Some of them are fairly good; some of them are not; and you can't know whether the content they're presenting is accurate. Some of them may cause you more confusion than help, so I recommend that you spend some hard-earned money and pay for a course that is taught by professionals. This expense will pay for itself in the long run because it will save you time and aggravation.

Following are some general study tips to help you prepare for the test:

>> **Target the testing sections.** In this chapter, you explore the sections of content the test covers. Work on one section at a time and try to master the skills and content in that section before you move on. When you do move on, keep reviewing the previous sections as well. Doing so reinforces what you've already learned as you add new content.

>> **Spend more time on sectional charts.** The sectional chart questions test your ability to interpret airspace and overlapping airspaces and the rules around them. You get chart questions asking about communications frequencies. These are all separate skills combined into one chart. Work on quiz questions that test your ability to find what you need on these charts. The more you practice, the easier those questions get. It takes some time, so try not to get frustrated as you go.

>> **Don't cram.** It takes time to truly learn the Part 107 test content. Take your time and try not to have unreasonable expectations about the future exam day. Some people learn the content faster than others. Just give it the time you need. Breathe, relax, build in breaks for your brain, and take longer than cramming gives you to prepare. In the end, you'll be happier with the results.

>> **Learn concepts.** Some of the Part 107 exam is conceptual. The FAA wants you to understand a drone piloting concept and then apply that concept to different piloting situations. Try to step back from the details occasionally and think about the big issue. Why is the information important, and why does the FAA want you to know it?

>> **Memorize facts.** Some aspects of the exam require you to memorize information. You need to memorize facts about airspaces, radio frequencies, and other details. Identify items along the way that you have to commit to memory and work on memorizing them.

>> **Use flashcards.** For facts and details that you need to memorize, old-school flashcards still work great. Practice with them. Have someone else ask you questions so that you have to verbally answer. These tactics help your brain remember.

>> **Draw pictures.** Visual aids help your brain to remember and make connections among airspaces, runway patterns, and other elements. Practice drawing rough sketches over and over. When I took my Part 107 test, I took the piece of scratch paper provided at the testing center and drew some items I had trouble remembering — before I started the test. I also wrote down a few things I tended to stumble over. Then I used the scratch paper to refer to my drawings and notes as I took the test. This tactic may work well for you. Don't just memorize words: Memorize pictures!

>> **Practice visually and verbally.** As you practice for the exam, don't simply read silently or look at drawings. Talk about the information. Read it out loud. Some people are better visual learners; some are better auditory learners. Combine the two approaches so that you see and hear the information. Run flash-card drills but say them out loud as practice. Doing so reinforces the content and helps you learn faster.

Registering for the Part 107 Exam

When the day to register for the exam finally arrives, you'll probably be both excited and nervous. That's normal, so don't stress about it. If you've studied well, you'll pass your Part 107 exam and have your certification soon. Exciting!

When you're ready to schedule your exam, follow these steps:

1. Go to https://www.faa.gov/uas/commercial_operators/become_a_drone_pilot.

2. **Get an FAA Tracking Number (FTN) by creating an Integrated Airman Certification and Rating Application (IACRA) profile.**

   You need to create this profile before you register for your test.

3. **Schedule your test appointment with an FFA-approved testing center.**

   You can find a testing center based on your location. You may have a few to pick from, but know this: Some offer regularly scheduled tests, and some do not. You may need to check a few of them to find a time and day that work best for your schedule. You can also schedule the appointment by finding the testing center at the link in Step 1 in this list. Don't forget that when you take the test, you need to bring a government-issued photo ID, such as a driver's license or passport, with you.

4. **Take and pass your test!**

   Arrive at the testing center at least 15 minutes before your scheduled test time.

After you pass your test, you need to complete the FAA Form 8710-13 for the remote pilot certificate. You use your previously created IACRA account to complete this form. Follow these steps:

1. **Log in to your IACRA account and choose Start New Application.**

2. **Choose Pilot for the application type and Remote Pilot for the certifications; then just follow the prompts.**

   You'll need the 17-digit Knowledge Test Exam ID, which may take up to 48 hours to appear from your test date.

3. **Sign and submit the application.**

   You'll get a conformation email after the system runs a TSA security background check on you.

4. **Print the email to use it as your temporary certificate until you receive your permanent certificate in the mail.**

5. **When that day comes, celebrate!**

TIP

Always have your FAA certification physically with you when you fly your drone.

REMEMBER

After you receive your certification, you need to complete some online recurrent training every two years to keep your certification current. You can find out more at https://www.faa.gov/uas/commercial_operators/become_a_drone_pilot.

Chapter **6**

Following Flight Regulations

The Federal Aviation Administration (FAA) has numerous regulations that apply to you, as a drone pilot, about your potential flights. Sometimes it's easy to think, "There are too many rules!" Sure, it may feel that way, but remember, the FAA's goal is to keep the skies safe. The sky is full of all kinds of traffic, and the flight regulations are designed to help you avoid an accident. Accidents lead to property damage, and in the worst-case scenario, the loss of life. It's important to realize that the regulations exist for everyone's safety.

This chapter covers some of these regulations and how different rules govern different kinds of flight scenarios. In this chapter, you find out about FAA regulations concerning height, speed, visibility, visual line of sight (VLS), and Remote ID. (I cover more regulations in Chapter 7.)

TIP

The Part 107 exam also expects you to know these regulations and apply them in different piloting scenarios.

Understanding Maximum Height

REMEMBER

The maximum height a drone can legally fly is 400 feet. There, that was an easy section! Well, the topic has a bit more to it than that, of course. There two exceptions to this requirement. And you may well ask what 400 feet means — 400 feet from what, exactly? Those are good questions, and the next sections supply the answers.

Understanding MSL and AGL

The concept of height for flying is governed by two different measurements for altitude. The first is *mean sea level (MSL)*, which is the average height that you would fly if you took off from sea level. Different flight levels affect your drone's performance because the atmosphere is different at different altitudes, which is why this measurement matters.

If you take off from the beach and fly to 300 feet, you're clearly at 300 feet MSL. However, if you take off from a 400-foot MSL mountaintop beside the beach and fly to 300 feet from there, now you're flying at 700 feet MSL. See Figure 6-1.

FIGURE 6-1:
Mean sea
level (MSL).

However, MSL isn't used much for drone height regulations. Instead, the relevant measurement comes from *above ground level (AGL)*, which means that the drone's maximum height of 400 starts at the ground level (regardless of the ground's elevation) and extends up to 400 feet. So whether you take off from the beach or a mountaintop, your maximum allowed height from your takeoff point is the same: 400 feet AGL, as shown in Figure 6-2.

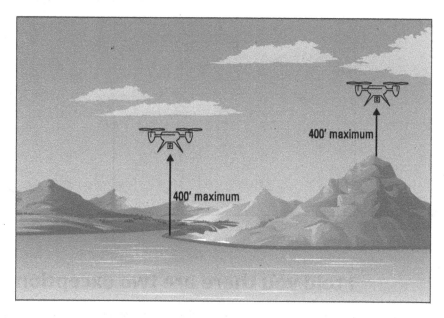

FIGURE 6-2:
Above ground
level (AGL).

REMEMBER

Understanding MSL and AGL is important, first because of how different elevations affect the performance of your drone's flight and battery life. The performance of your drone at 400 feet depends on where you take off from. If you take off from Galveston, Texas, at the beach, your AGL and MSL are the same: They're both 400 feet because you took off from sea level. But taking off from Denver, Colorado, is a different situation. Denver is about 5,200 feet above sea level (which is why it's called the Mile High City), so if you fly 400 feet above Denver, you're flying at roughly 5,600 MSL. You're legal because you're at 400 feet AGL, but in terms of performance, you're in thin air at a mile high, and your drone will therefore have to work harder. Understanding the difference between MSL and AGL helps you understand performance issues at various altitudes.

REMEMBER

A second reason to know about MSL versus AGL relates to sectional charts of airspace (see Chapters 11 and 12 for more information about sectional charts). Sectional charts basically represent a "section" of airspace with information about what is in that airspace from the ground up. Sectional charts label various structures and tell you the height of those structures. On sectional charts, all altitudes are noted in MSL. If there is an exception for AGL, the sectional chart displays that number in parentheses. For example, Figure 6-3 shows a sectional chart with several towers. Beside the tower noted in the image are two numbers. The number

2519 means that the top of the tower is 2,519 feet MSL. The number 492 in parentheses under it means that the top of the tower is 492 AGL.

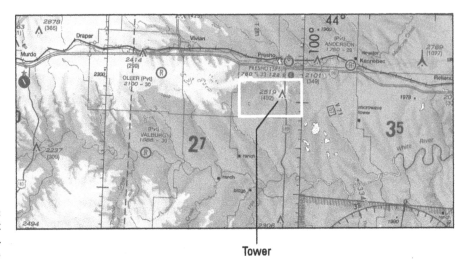

Tower

I told you there are two exceptions

REMEMBER

As stated previously, 400 feet is the maximum height you can fly your drone. However, the FAA allows two exceptions to this rule. First, if you are flying next to a structure, you may fly up to 400 feet above the topmost part of the structure while maintaining a radius around the structure of 400 feet as well. The FAA doesn't directly define what a "structure" is.

To apply this idea in practical terms, say you've been hired to inspect a tower that is 562 feet AGL. You need to get to the top and above it to take photos. The FAA rule states that you can fly up to 400 feet above the tower's uppermost part as long as you stay within a 400-foot radius of the tower. In this case, you could fly 962 feet AGL (the 562-foot tower plus an additional 400 feet), as shown in Figure 6-4.

TIP

However, note that this rule isn't always true because you have to consider the airspace. For example, you could be hired to inspect a tower that is 562 feet tall, but it exists in an airspace that restricts your maximum height. In this case, you may be able to get permission to fly higher than allowed. See Chapter 7 for more about obtaining airspace restriction waivers.

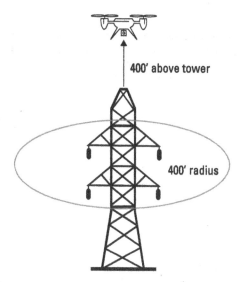

400' above tower

400' radius

FIGURE 6-4:
Sectional chart
MSL and AGL
notations.

The second exception to the 400-foot maximum concerns safety. As a drone pilot, you must yield right-of-way to any other aircraft, and if you need to break the 400-foot maximum to safely yield right of way, you may do so. For example, say you're flying your drone at 350 feet and suddenly see a helicopter near you. If the safest way to get away from the helicopter is to fly up and beyond 400 feet, do it. In these cases, you should always try to reduce your drone's elevation to avoid another aircraft, but if you can't, then fly up!

Exploring Speed and Weather Visibility

The FAA's rule for speed is straightforward: You cannot fly your drone faster than 87 knots, or 100 miles per hour. That's probably not an issue because your drone probably isn't likely to fly that fast anyway. However, some commercial drones can, so that's why this rule is in place.

REMEMBER

The rule concerning weather visibility is where things get a bit tricky. The rule states that you must be able to see for three miles from the location of the control station, meaning where you take off and control your drone using the RC (just think of the control station as yourself). In other words, the FAA wants you flying only in clear conditions because poor weather and fog pose a flight risk. See Chapter 14 for more information about cloud cover and weather conditions.

In addition, when clouds are overhead, you must fly 500 feet below the clouds. This rule applies if you're flying directly under some clouds (think heavy, dark cloud cover).

However, if you're flying at the same level as the clouds (such as in a high elevation area, like mountains), you must move away from the clouds and keep a 2,000-foot horizontal distance between your drone and them. Figure 6-5 illustrates these rules.

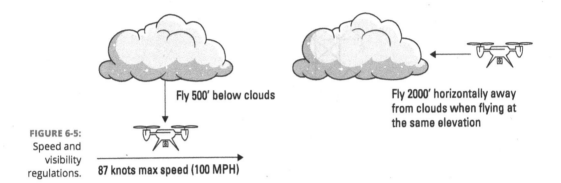

FIGURE 6-5: Speed and visibility regulations.

Fly 500' below clouds

Fly 2000' horizontally away from clouds when flying at the same elevation

87 knots max speed (100 MPH)

Keeping a Visual Line of Sight (VLOS)

As I mention elsewhere in the book, the FAA expects you to keep a visual line of sight (VLOS) of your drone as you fly. This means that you must be able to physically see your drone when you fly. Your direct line of sight can't be obstructed by buildings, trees, hilltops — anything else. As you might guess, this rule exists for safety purposes.

Basically, you want to do three things: Keep an eye on your drone's physical location, keep an eye on the RC, and have another eye scanning the airspace for obstructions or other aircraft. Wait . . . that's three eyes, but you get the picture.

The VLOS rule comes with a few sub-rules as well:

>> **You can use a visual observer.** A *visual observer* is another person located near you and in communication with you who helps you keep VLOS as you fly and use the drone's RC.

>> **A first-person view (FPV) drone requires an observer.** If you're using FPV headsets or similar, you must have a person with you who can physically see the drone without any visual aid, such as binoculars. Any FPV headset has a built-in screen, which makes flying feel like you're physically in the air with your drone. It's fun, but it's not a replacement for VLOS because you're seeing the world from the drone's camera instead of your eyes.

WHAT IF YOUR DRONE IS TOO SMALL TO SEE?

You may be thinking, "My drone is white and weighs less than 250 grams — it's small! How am I supposed to see it against the sky if I have any altitude at all?" That's a true concern. After all, many people use small drones to stay under the 250-gram weight limit (explained in Chapter 1), and they can be hard to see in the air. The FAA doesn't give you an exception for this, so you might consider taking a few actions to make your drone easier to see when it's in the air:

>> **The visual observer doesn't have to be Part 107 certified.** Although the visual observer doesn't have to have a Part 107 certification, the observer should know what you want them to do, which is keep a line of sight on the drone while scanning for any possible airspace obstructions or problems.

TIP

Keep in mind that VLOS requires having a line of sight to the drone without visual aids, such as binoculars. This rule is true for you as the pilot but also for the visual observer. You can wear glasses or contacts, but you can't use binoculars.

>> **Use reflective tape or stickers.** Consider adding some reflective tape or stickers to your drone that makes it stand out against the sky. (Orange is a good color.) Make sure that you don't put tape or stickers over any sensors if your drone is equipped with them.

>> **Add lights.** You can buy inexpensive light kits that place strobing lights on your drone so that it's easier to see in the sky.

>> **Fly with the sun behind you.** The sun creates glare, and if you have glare, you can't see your drone. Position yourself so that the sun is behind you to eliminate most of the glare.

Flying During Twilight

No, we're not talking about vampires here. In the United States, evening twilight is the period time from sunset to thirty minutes after sunset. Morning twilight is the thirty-minute period before sunrise.

TIP

You can fly your drone during twilight hours legally as long as you adhere to one rule: Your drone must be equipped with anti-collision strobe lighting that is visible for three miles. You can easily add this lighting to your drone if you need to fly during twilight hours. Although many lighting options are available, one of the most popular is Lume Cube. Check it out at `https://lumecube.com/`.

Complying with Remote Identification

Remote Identification (Remote ID) is a newer FAA regulation that requires drones that must be registered (they're over the 250-gram weight limit) to have the ability to broadcast location and identification information to other devices around the drone as it's flying. The purpose is safety. The higher the number of drones in the airspace, the more problems there can be, so think of Remote ID as a safety feature that identifies your drone. It's sort of like a digital license plate.

Remote ID can help the FAA or law enforcement locate a control station (you) of a drone that is flying in an illegal or unsafe manner or flying in a location that's not legal because of airspace or other restrictions.

If your drone weighs more than 250 grams, it needs to be Remote ID compliant. You have two ways to meet this compliance requirement: Standard Remote ID or Remote ID Broadcast Module. The following sections explain each of these options.

Standard Remote ID

A Standard Remote ID drone has the broadcast module built into it, so when you buy the drone, it's already Remote ID compliant and ready to fly. Most newer drones (since 2022) are Standard Remote ID, or they are at least Remote ID ready with a firmware update. You can take a look at your drone's make and model compliance at `https://uasdoc.faa.gov/listDocs`.

As shown in Figure 6-6, a Standard Remote ID drone broadcasts several pieces of information during flight, such as the drone's ID, location, altitude, velocity, location of the control station (you), and so forth.

TIP

If you're shopping for a new drone, you may notice that many of them say "FAA Compliant" or something similar. This notice typically means that the drone has a built-in Remote ID, but don't assume; check it out first before you part with your hard-earned money.

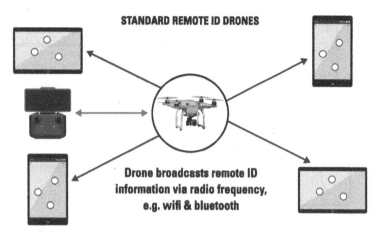

DRONE REMOTE IDENTIFICATION

STANDARD REMOTE ID DRONES

Drone broadcasts remote ID
information via radio frequency,
e.g. wifi & bluetooth

◆ Remote ID capability is built into the drone
◆ From takeoff to shutdown, drone broadcasts:
   ◆ Drone ID
   ◆ Drone location and altitude
   ◆ Drone velocity
   ◆ Control station location and elevation
   ◆ Time mark
   ◆ Emergency status

FIGURE 6-6:
Standard
Remote ID.

Remote ID broadcast module

If you have a drone that doesn't have a Remote ID built into it, such as an older model, you can add a Remote ID broadcast module to the drone. The broadcast module is a small device that you place on the outside of the drone with some kind of adhesive tape or fastener product. After you add the module, it broadcasts the same information as a Standard Remote ID, as shown in Figure 6-7.

TIP

Remote ID broadcast modules are manufacturer and even drone specific, so check out your drone manufacturer's website for more information about the broadcast module you need and where you can purchase it. Remote ID modules also need to be FAA approved. You can find out more at https://www.faa.gov/uas/getting_started/remote_id.

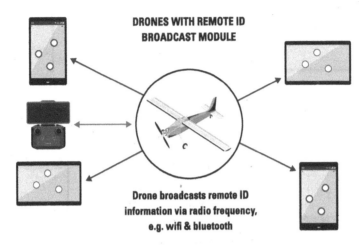

DRONE REMOTE IDENTIFICATION

DRONES WITH REMOTE ID
BROADCAST MODULE

Drone broadcasts remote ID
information via radio frequency,
e.g. wifi & bluetooth

◆ Remote ID capability through module attached to drone
◆ Limited to visual line of sight operations
◆ From takeoff to shutdown, drone broadcasts:
    ◆ Drone ID
    ◆ Drone location and altitude
    ◆ Drone velocity
    ◆ Takeoff location and elevation
    ◆ Time mark

FIGURE 6-7:
Remote ID
broadcast
module.

FRIA exception

There are some places where you can fly a drone without a Remote ID. This kind of place is called an FAA-Recognized Identification Area (FRIA). Both the drone and the pilot must be located inside the FRIA for the exception to apply, and if you have a drone that has a Remote ID, it still must be used inside the FRIA, as shown in Figure 6-8.

TIP

FRIAs are often established with the FAA through various drone flight clubs and such, so to find out whether you're in an FRIA, check out the FRIA locations on the web at https://faa.maps.arcgis.com/apps/webappviewer/index.html?id=8f 274117010f4eb1a50f64c1719be12b.

DRONES WITHOUT REMOTE ID

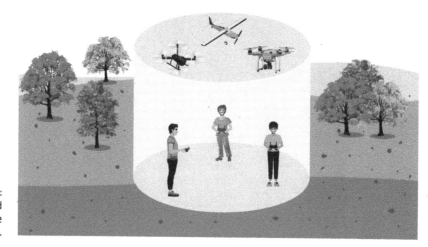

FIGURE 6-8:
The drone and
pilot must be
inside of the FRIA.

» **Flying over property and people**

» **Flying from a moving vehicle**

» **Reporting an accident to the FAA**

» **Logging your flights**

» **Getting flight authorization waivers**

» **Understanding the effects of alcohol, drugs, and medications**

Chapter **7**

Following More Flight Regulations

Chapter 6 explores some flight regulations, and this chapter has even more! As you can clearly see, the Federal Aviation Administration (FAA) does have numerous flight regulations for drone pilots to follow. Luckily, most of them aren't difficult, and they're all designed to keep people and property safe.

In this chapter, you explore the rules pertaining to flying over property, people, and moving vehicles. I also cover regulations for reporting an accident, logging your flights, and obtaining authorization waivers for restricted airspace. And finally, the chapter delves a bit into the impact of drugs, alcohol, and prescription medications on the safe operation of a drone. All the regulations in this chapter greatly impact your day-to-day work with a drone, especially if you move into some kind of piloting career. (If you want more information about drone piloting careers, be sure to check out Chapter 19.)

Flying Over Property and People

Of all the FAA's Unmanned Aircraft System (UAS) regulations, the rules about flying over property or directly over people have the biggest impact on your day-to-day operations. It's also the aspect of drone flying that causes complaints from other people. Chances are, you'll have to explain these rules to someone at some point (perhaps even a law enforcement officer), so it's a good idea to really understand these principles and be able to explain them.

Flying over property

During many drone flights, you're likely to fly over someone's private property. *Property* in this context means someone's land or residence — that is, land or a house that another person owns. It could range from a large farm to a residential house in a suburb.

Confusion abounds about drone pilots flying over another person's private property. The rule isn't confusing; it's just that most people don't understand it. Property owners do not own the airspace over their property; the government does. Therefore, the FAA makes the rules about flying over personal property. So here's the rule: You can fly your drone over someone's property at any altitude you like, as long as you don't fly higher than 400 feet. That's the law.

REMEMBER

Along with that law, you must be free to fly in the airspace where the property exists (see Chapter 9), or if you aren't, you must have a waiver to fly in the airspace. (See "Getting Airspace Authorization or Waivers," later in this chapter, for more details.)

Yet another complication with this law is that although you can fly over someone's property, you can't pause and photograph or video someone enjoying their backyard pool or yard without their permission. That's invasion of privacy. You're free to fly over the property, but you're not free to stop, hover, or invade someone's privacy. Figure 7-1 illustrates these principles.

Think about it. Depending on where you live, commercial airplanes may be flying over your property at any moment, and you don't think anything about it because the planes are so high up. But a drone flies lower, and people often think, "You can't fly over my house!"

In fact, many drone pilots flying for fun or work have encountered angry property owners who complain about privacy or even call the police. This reaction occurs because most people don't understand who owns the airspace above that property. If you run into trouble in this kind of situation, here's what you should do:

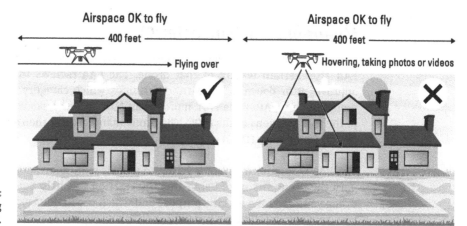

Airspace OK to fly   Airspace OK to fly

400 feet   400 feet

Flying over   Hovering, taking photos or videos

FIGURE 7-1:
Rules for flying over property.

>> **Kindly explain.** Be kind and friendly and explain to the property owner that you are flying in FAA airspace and have permission to do so. If you have your Part 107 license, kindly throw in that you're an FAA-certified UAS pilot.

>> **Kindly reassure.** Reassure the property owner that you are passing over the property only. You're not observing it or taking any photos or videos.

>> **Be prepared for law enforcement.** In the case of an irate property owner who calls the police, be prepared to show your TRUST certificate or Part 107 license. You may have to explain the rule to the police officer because many of them will not know the FAA rules. Police officers are generally not trained in FAA drone piloting rules, and after all, that's not their job. Be kind, calm, and professional. In most cases, when a law enforcement officer sees your license and hears you professionally explain the rule, that's all you'll need to resolve the problem.

Flying over people

Flying your drone over property is fairly straightforward in terms of the rules. Flying your drone over people gets more complicated, however, resulting in a lot of confusion. The rules are more complicated because of the much greater risk of hurting a person in the case of a failure and crash. In this section, I make these rules easy to understand by showing you how the several different concepts involved all work together.

Means of compliance (MOC)

A means of compliance (MOC) is a document that a drone company provides to the FAA after certain testing of that drone. The FAA reviews the MOC and either approves it or doesn't. The MOC determines which category the specific drone model fits into. After the FAA approves an MOC, drone models are manufactured to the specifications of the MOC. The drone manufacturer then submits a declaration of compliance (DOC) to the FAA.

TIP

The FAA maintains a database of drone models with information about the category the drone fits into. Using this information can help you determine which drone you need for a specific purpose and which restrictions exist, including what you can do with the drone concerning flying over people.

You can check out this information at `https://uasdoc.faa.gov/listDocs`. On the website, you can search for a specific drone and view the information about it. For example, Figure 7-2 shows the DOC for AgEagle's eBee X drone. From the information, you can see that the DOC is accepted by the FAA, and that this drone is a Category 3 drone. Of course, that begs the question, "What are the categories?" I get to that shortly, in the upcoming "Drone categories" section.

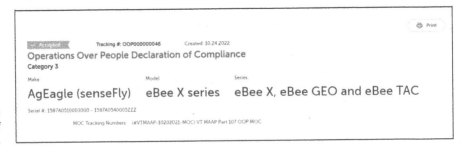

FIGURE 7-2:
A declaration of compliance (DOC).

Understanding what "flying over people" means to the FAA

Flying over people means that you fly your drone above a human being or even in the vicinity where a crash could cause a person to be injured. The phrase applies to situations like flying around a neighborhood or park where people may be outdoors enjoying the day.

Flying over people also includes outdoor gatherings or assemblies, such as an event at a park, a concert, a Little League game, or something like that.

To the FAA, "flying over people" applies to individuals as well as assemblies of people gathered outdoors.

TIP

You also need to understand the concept of "sustained flight" concerning people. *Sustained flight* refers to intentionally flying your drone over a crowd of people, moving back and forth over that crowd, or circling that crowd (often for the purpose of taking photos or video). If you just pass by or somewhat over a crowd while flying your drone to a different location, that's not considered a sustained flight.

Drone categories

The FAA classifies drones by various categories, with each category determining what you're allowed to do with your drone. These categories are based on the drone model's Declaration of Compliance (DOC) that the manufacturer submits to the FAA, as described earlier in the chapter. Table 7-1 describes the categories.

TABLE 7-1   **Drone Categories**

Category	Explanation
1	Drones in this category may fly over people. The drone must weigh less than 250 grams and contain no exposed rotating parts that could cause a skin laceration should the drone crash and strike a person. Additionally, you can't operate any drone in Category 1 in sustained flights over crowds unless the drone is Remote ID compliant (see Chapter 6 for more details on Remote ID).
2	Drones that weigh more than 250 grams can fly over people. These drones provide performance-based data to the FAA, although they may not have an *airworthiness certificate,* which is yet another FAA certification that drone manufacturers may request depending on the weight of the drone and the safety features of the drone. Drones in this category may not perform sustained flights over crowds without being Remote ID compliant.
3	This category also covers drones that weigh more than 250 grams, have performance-based data, and don't have an airworthiness certificate. However, drones in this category may not fly over assemblies of people and may fly over people at all only if the operation is within or over a closed or restricted-access location and the people are notified that a drone is flying over them. Also, the drone may not perform any sustained flight over a person unless that person is participating in the flight operation or is located under a covered structure or in a vehicle, both of which provide protection from the drone should it crash.
4	Drones in this category have an airworthiness certificate and may fly over people, following the same rules as Category 2.

The drone categories may seem a bit confusing, but keep in mind that a lot of the restrictions relate to the drone's weight and airworthiness certificate. In other words, based on the drone's construction and weight, if the drone crashed into a person, what kind of harm would it likely cause? But keep in mind that there's a balancing act among potential harm, airworthiness, and even remote ID. The good news is that most drones you'll fly fall into Category 1 or 2. Commercial drones may fall into Category 3 or 4.

Another concern relating to possible harm is that of skin laceration. Most drone propellers move fast enough to lacerate someone's skin if the drone hits a person during a crash. (The exceptions are some very small toy drones.) To be FAA compliant in Category 1 or 2, drones can't have any exposed moving parts, which means that you need to use propeller guards. Propeller guards are just pieces of plastic that are drone specific and attach to the drone to cover the propellers, as shown in Figure 7-3. They're inexpensive and you can find them anywhere that sells drones. However, note that putting propeller guards on a drone weighing less than 250 grams may make it exceed the 250-gram limit, so that's something to keep in mind from a legal standpoint if you're flying over people.

FIGURE 7-3:
Drone
propeller
guards.

yurakrasil/Adobe Stock Photos

If you take the Part 107 exam, you may see a question or two about a drone's category and the rules for flying over people. It's a good idea to have the categories table memorized!

Flying over large groups of people

Although you can fly a drone over assemblies of people based on the rules of the drone's category, note that this rule isn't unlimited. In fact, no drone is allowed to fly within three nautical miles of a stadium that can seat 30,000 people or more. In other words, major sporting events and even potentially large outdoor concerts are strictly prohibited.

You can't fly at all near these events during the event or one hour before or after the event. Doing so can result in arrest and potentially serious fines, so keep away from large-scale assemblies of people.

Flying from a Moving Vehicle

Sometimes you may feel the need to fly your drone over a large piece of property. The best way to stay close to your drone during this flight is to fly from an automobile traveling down the road as the drone flies over the property. You have someone to drive the automobile so that you can fly from the passenger's seat. But is this legal?

TIP

Generally, the FAA prohibits flights from any moving land- or water-based vehicle (think car or boat). However, the FAA does make an exception if you're flying in a sparsely populated area. So, in the preceding example, if the property is a big piece of land with few people around, it's okay to fly from the automobile in this case.

However, the vehicle you use must be a land vehicle. You can never fly your drone if you're in a helicopter or airplane. You must be on the ground.

REMEMBER

In any flight scenario (such as flying from a moving vehicle), you must keep your drone flying under 400 feet and be allowed to fly in the airspace.

Reporting an Accident to the FAA

In a perfect world, this section of the chapter wouldn't be needed. Unfortunately, sometimes accidents happen and drones crash. In certain circumstances, you must report an accident to the FAA. According to the FAA, you must report an accident if

>> **A serious injury occurs.** A series injury means that your drone hits a person and causes a loss of consciousness, a broken bone, a skin laceration requiring sutures or any situation in which the person needs medical help. Scratches and "band aid" types of injuries do not need to be reported.

>> **Significant property damage results.** If the drone accident causes more than $500 in property damage (such as to a person's car, house, or other property), you must report it to the FAA.

Either of these scenarios require you to report the accident to the FAA within ten calendar days of the accident.

To report an accident, just go to the FAA Drone Zone at https://faadronezone-access.faa.gov/#/ and log in to your account. You'll see the option to report an accident. The form asks for your contact information and information about the drone, and you need to enter details about the accident. Remember, reporting accidents is the law, so don't skip it if one ever occurs.

Logging Your Flights and Maintenance

Here's the thing: The FAA governs drone flights, and an officer of the FAA can ask you at anytime for information about your flights, inspect your drone, and even ask for flight logs and records of your drone's maintenance. This event is unlikely, but if you were involved in an accident or were working in a high-risk location for a job, you could encounter this request.

So you want keep flight logs as well as any notes or logs about drone maintenance, such as repairs, flight time on a set of propellers, and so forth.

You can log your flights manually, or your drone may already be doing it for you. Many drones log flight data, and those log files are often stored on the RC's memory card. You need to check your drone's documentation for specifics. If your drone is automatically logging your flights, you don't have to worry about doing it.

If your drone doesn't automatically log flight data, just get a notebook and log each flight's location, time, date, and flight time. For enterprise operations, many individuals and companies use additional software and tools to keep track of things. Drone Log Book at https://www.dronelogbook.com/hp/1/index.html is an example of one you can check out.

Aside from flight logging, be sure to keep up with maintenance, battery lifespans, and other mechanical repairs or problems.

To know when you should perform maintenance, including how often to replace the propellers or the drone's batteries, be sure to check out the drone manufacturer's suggested schedule of maintenance activities. As you log your drone flights, you can keep aware, for example, of how many flight hours are on a set of propellers and know when the manufacturer suggests changing them for a new set.

REMEMBER

In the end, a key to avoiding accidents is to prevent them. On-time maintenance of your drone's components can be a big factor. The FAA expects you to keep track of your drone's maintenance, so it's best to get a plan in place and follow the manufacturer's guidelines.

Obtaining Airspace Authorization or Waivers

You can explore important considerations relating to airspace in Part 3 of this book, but before you do, be aware that all airspace except G has restrictions. Sometimes you may need to fly in a restricted area for a flight operation, and to do so, you need to request authorization from the FAA. There are two kinds of authorizations that can be given by the FAA: Airspace Authorization or an LAANC waiver.

Getting airspace authorization

When you say airspace authorization, think airport (or an area near an airport). These restricted areas are operated by Air Traffic Control (ATC) for that specific airport or area. You can't fly in these areas for obvious reasons, but sometimes you really do need to. For example, you may get a drone piloting job that requires flying near an airport for a brief period of time.

Before you can enter restricted airspace, you need to apply for airspace authorization, which you can do through your FAA Drone Zone account. But how do you know whether you need airspace authorization? This is where understanding airspace and reading sectional charts (which Part 3, especially Chapter 11, explains) come into play. Sectional charts contain airspace information around the country so that you can determine a geographic area's airspace and whether you can fly in it.

Receiving airspace authorization through your Drone Zone account can take up to 90 days. Not so practical.

Fortunately, however, you can in most cases get instant approval for most controlled airspace areas. Unless you need fly next to a big airport like Atlanta or DFW, the odds are good that you can get approval right away by requesting a LAANC authorization, described in the next section.

Getting LAANC authorization

The Low Altitude Authorization and Notification Capability (LAANC) is a system in which drone manufacturers and other companies work with the FAA to provide flight access in restricted airspace. You can use an app on your phone to request authorization, and it's often granted instantly.

TIP

Free apps that partner with the FAA are available to help you quickly find out what airspace you need to fly in. Those apps also enable you to request LAANC authorization. For example, Figure 7-4 show the AutoPylot app used in a particular location. The app indicates that you fly here with LAANC approval. You click Continue to request approval for that specific area, and the result often arrive immediately or within a few minutes. Sometimes the result tells you that approval is granted, but with an altitude restriction under the 400-foot maximum. Your results depend on the location and what may be going on in the skies at that moment.

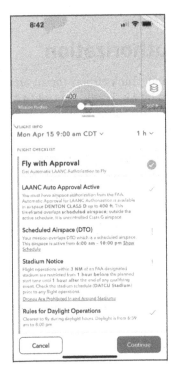

FIGURE 7-4:
Getting LAANC authorization using AutoPylot.

The FAA maintains a web page with up-to-date information about LAANC as well as which apps you can download and use for approval. Check it out at https://www.faa.gov/uas/getting_started/laanc.

YOU CAN'T FLY IN NATIONAL PARKS

National parks would be great places to fly a drone because the scenery is often amazing. Unfortunately, you can't. Flying a drone in any national park in the United States is not legal. In fact, most state parks also prohibit drones. (Exceptions exist, so always check the rules at a particular state park.) In some instances, you can get access to fly in a national park, but that authorization is rarely granted. Check out the national park's website for details or possible exceptions, but otherwise, don't fly in a national park. You can face some stiff fines for doing so.

Understanding the Effects of Alcohol and Drugs

Alcohol, drugs, and prescription medications can have serious effects on a pilot's ability to fly a drone safely. Even a small amount of alcohol can affect your muscle movements and eyesight when you operate the RC. Alcohol and drugs can greatly impair your response time and ability to make quick flight decisions.

REMEMBER

The FAA gives us this rule: Your blood alcohol content must be less than 0.04, and eight hours must pass after you have a drink before you can fly your drone.

Say you have dinner and a glass of wine with friends and then want to show your friends your new drone in flight. That's a big no-no with the FAA. Don't do it!

Various drugs affect the body as well. The word "drug" tends to make people think of illegal drugs (which are absolutely prohibited in aviation), but many prescription medications have potentially negative effects on drone operation as well. For example, even some antibiotics can affect your balance!

To find out whether a prescription medication can prevent you from flying safely, first look at the side effects of that medication and ask your pharmacist specific questions. That's the best place to start.

Otherwise, the FAA has Aviation Medical Examiners (AMEs) who can probably answer questions you may have. You can find an AME in your area at this link: https://www.faa.gov/pilots/amelocator.

WARNING

Flying your drone under the influence of alcohol or drugs is a serious offense that can result in accidents as well as the loss of your drone piloting license.

ADDITIONAL FLIGHT-READINESS ISSUES

Aside from alcohol and drugs, several other factors can affect a pilot's ability to safely fly a drone. Stress, low blood sugar, dehydration, or lack of sleep can all impact the safety of a flight. If you're nervous about a particular flight, you can experience hyperventilation, which is caused by too much oxygen due to shallow breathing. Hyperventilation can result in dizziness or feelings of faintness. All these issues pose potential flight risks. Before you fly, ask yourself, "Am I mentally and physically fit to fly at this time?" If you're in doubt, not feeling well, or some stressor is taking all of your mental energy, it's always best to stay grounded until you feel better.

IN THIS CHAPTER

» **Exploring FAA regulations for night flights**

» **Understanding drone limitations**

» **Considering eyesight limitations**

» **Understanding night-flight visual problems and illusions**

» **Developing night-flight best practices**

Chapter **8**

Flying at Night

As a drone pilot, you may sometimes want to fly at night for fun, but the odds are good that you may also need to perform some night flights for a piloting job. Perhaps a client needs some night photos of a building or area of a city, or some kind of emergency inspection is needed at night. You may even be involved in nighttime search and rescue operations.

You may think that flying at night is the same as flying during the day — just darker. However, night flying involves rules and safety considerations that differ from daytime flights.

This chapter tells you what you need to know about Federal Aviation Administration (FAA) nighttime flight regulations and how low lighting affects your photos and videos. You discover how and why your eyes work differently in darkness, along with strategies for dealing with the various optical illusions and limitations that can create night-flight challenges.

REMEMBER

As a responsible Unmanned Aircraft System (UAS) pilot, you want to have the right attitude. Overconfidence often leads to accidents, so it's important to say to yourself, "If I fly at night, there is more risk and a greater potential for an accident." Having the right attitude from the start can help you avoid potential problems with night flights.

Exploring FAA Regulations for Night Flights

Rules come and go, and they change with the times. That's true for flying your drone at night. In the past, night flights were prohibited without a waiver, but now you can fly your drone at night without any special permission, whether you have a Part 107 certification or not.

REMEMBER

Always keep in mind that any FAA or police officer can ask to see your Part 107 license or TRUST certificate anytime you're flying. Keep that license with you at all times, both during daylight hours and at night.

TIP

Flying during civil twilight hours has the same rules as flying at night. Check out Chapter 6 for details.

Fortunately, the rules for flying at night are pretty simple and straightforward, as explained in the following sections.

Flying over people and moving vehicles at night

If you have your Part 107 license, you can fly over people and moving vehicles at night without any special permission. However, it's important to note that the same rules for flying over people, vehicles, and even property still apply whether you're flying during the daytime, twilight, or at night. Make sure that you understand these rules (covered in Chapter 7) and comply with them.

Airspace and LAANC waivers

In the past, night flights required special waivers, but as of 2021, the FAA rules governing air space and LAANC waivers are the same for night flights as for day flights. This means that airspace rules are the same 24/7. You can't fly in restricted airspace without an LAANC waiver, and if you're flying very close to an airport, you likely need permission to fly from air traffic control (ATC) for that airport or region. To find out more about obtaining LAANC and ATC waivers and permission, see Chapter 7.

REMEMBER

Also, keep in mind that you can't fly higher than 400 feet without a waiver anytime, night or day. So, in a nutshell, your FAA flight rules are the same regardless of when you fly (except for lighting, which I cover next). That's a good thing — after all, who wants to learn a second set of rules!

Anti-collision lighting

The only specific rule for night flights is that your drone must be equipped with anti-collision lighting. This is the same rule that governs civil twilight flights.

REMEMBER

Sometimes people think, "My drone already has lights, so I'm good to go." Not so fast. The rule states that the lighting must be visible for three statute miles (which means a standard mile containing 5,280 feet), and the light must be blinking (a strobe light). Built-in drone lights don't meet the visibility and strobe requirements, and you can't just add some fun lights. You have to add actual anti-collision lighting.

You can choose among a variety of product options for adding anti-collision lighting to your drone. One of the more popular products is made by https://lumecube.com/. You may also want to check out https://www.firehouse technology.com/

Typically, an anti-collision light is attached to the top of the drone with adhesive tape, as shown in Figure 8-1. The light is small but bright, and that's what you need. Whatever product you purchase, make sure it's an FAA-compliant anti-collision light (meaning that it strobes and is visible for three statute miles).

FIGURE 8-1:
Anti-collision lighting.

Understanding Drone Limitations with Night Flights

Some drones are particularly designed for night or low-light performance. Law enforcement and search-and-rescue drones come to mind. These drones are outfitted with specific night-vision technology and other mapping features that enable them to see at night — even inside dark buildings. See Chapter 19 for more details about industry-specific drones.

But you're probably not flying one of those types of drones. Even if you fly a pro-level drone, unless it's specifically designed for dark environments, there are three limitations — pertaining to cameras, sensors, and batteries — to be aware of. Read on!

Camera limitations

Your drone may have a great 4K camera, but the odds are good that it's not designed for night photography or videography. In a nutshell, a camera works by interpreting light as it hits the camera's sensor. The other exposure controls, which are aperture, shutter speed, and ISO, impact how that light interacts with the sensor (such as with how much light, for how long, and how sensitive the camera sensor is; see Chapters 17 and 18 for details about exposure). Sure, you can take photos and videos at night, but unless the drone camera possesses specific night-vision features, the outcome probably won't be awesome.

If you have a pro-level drone with a very good camera system (such as the DJI Mavic 3 Pro), you can certainly capture some good images if some ambient light is available (think streetlights, house lights, the moon, and such). However, just know that most drone cameras aren't designed for night photography.

Here are some of the problems you can expect if you shoot photos and video at night:

>> **Graininess:** If it doesn't have enough light, the camera sensor will have trouble interpreting what it sees. This situation often leads to artifacts in the image, commonly called *grain*, as shown in Figure 8-2. If you turn up the camera's ISO setting to compensate for low light, you often get even more graininess, so it's a sort of vicious circle.

>> **Starburst effects:** If you adjust your camera's exposure settings to get more light at night, you often get a starburst effect if the lens sees a bright light, such as a streetlight. This effect occurs because the settings are configured for low light, but then you introduce a bright light into the frame. These starburst effects can ruin a photo or video (also shown in Figure 8-2).

>> **Focusing problems.** The drone's camera system must be able to see well to focus on subjects and objects in the frame. If it can't, you can wind up with somewhat blurry images and video.

>> **Motion blur:** A strategy to get more light to the sensor in dark environments is to slow down the shutter speed. Doing so allows light to interact with the camera sensor for a longer period of time. However, slower shutter speeds create motion blur, so if you're flying and shooting photos with a slower shutter speed, you get blurry images.

FIGURE 8-2: Graininess and starbursts are common night photography problems.

Cofefe/Getty Images

Sensor limitations

Drones that fall into the prosumer category and higher (see Chapter 1 for more on drone categories) use sensors for flights and obstacle avoidance. Sure, the drone uses GPS to know its location in the world, but its sensors help the drone understand what objects are around it — and where the ground is when landing. Those sensors need light to see effectively.

Concerning obstacle avoidance, the drone's sensors identify what is around it at all times during a flight. If the drone senses an obstacle, it can stop or bypass the obstacle (depending on the settings you choose). But, the drone's sensors are not as good as your eyesight. For that reason, if you're flying low through a group of trees, the sensor may not detect a small branch and may crash into it. The sensors are good, but they're not perfect. But take away most of the light, and what can you expect? Not much help! So don't depend on the sensors for obstacle avoidance during night flights.

REMEMBER

Most drones have a sensor on the bottom that helps the drone see the ground when it lands. If the drone can't see the ground, it may not land when you tell it to. Some drones are outfitted with a spotlight on the bottom of the drone that comes on automatically during landing to help the drone land. These models can detect low lighting and use the spotlight when it's needed without any additional help from you. That's another feature to consider if you're shopping for a new drone!

Also, numerous reports on the internet tell of drones behaving erratically in low-light conditions. Some drones handle low-light conditions better than others, but unfortunately, you just never quite know what you're going to get.

WARNING

Be aware that drone sensors don't work as well or at all during night flights, so you really have to be on guard and avoid relying on these sensor systems; they may not help you much.

Battery limitations

Although it may not be a big issue for you, battery limitations are worth mentioning. Depending on the time of year you're flying, the temperature may be significantly colder after dark, and colder temperatures mean poorer battery performance. LiPo batteries don't last as long as other types in cold weather.

Also worth considering is that anti-collision lighting may add weight to the drone, which can reduce battery life (depending on the size of the drone and batteries, of course). Also, if you need your drone to stay under the 250-gram threshold, anti-collision lighting can push it over the limit.

Overall, it's a good idea to keep in mind that night flying may lessen battery charge time. If you remember that, you can make better night-flight decisions concerning the time and distance you'll fly.

Considering Eyesight Limitations at Night

Your drone certainly has some limitations with night flights, but the truth is, you have some limitations, too! People often believe that they see well at night, but here's the truth: No, you don't. Human beings just don't see well in the dark.

Compared to animals, our nighttime eyesight is just terrible. We humans surround ourselves with all kinds of artificial light in dark environments and at night for one simple reason: so that we can see.

TIP

Although this may seem like an odd topic for a drone piloting book, it's important to understand your night-vision eyesight and its limitations because the FAA expects you to understand these principles. If you take the Part 107 exam, you will likely see a test question or two about your nighttime eyesight limitations, which the following sections explain,

Understanding the basics of your eyesight

Without going too deep into biology, this section provides a quick look at how your eyes work. Your eyes consist of the same basic components as a camera (see Figure 8-3). In fact, modern cameras are all built on the biological principles of your eyes, which is pretty cool when you think about it.

Your eyes' job is to pass light signals to your brain. Your brain interprets those signals so that you can see and make sense of the world. From your eyes, you get shapes, color, textures, and even depth perception. You can physically turn the page in this book because of your eyes' ability to interpret depth. Also, you can focus on both faraway and close objects (perhaps with some help from glasses, as I do). Our eyes are really quite amazing when you think about it.

When you look at an object, light travels through your eye and your brain interprets those signals. In fact, the images your eyes gather are inverted (upside down), but your brain flips those around so that you "see" the world as it is. Here's a quick look at how it works:

1. **The cornea bends the light.** The cornea is the clear, dome-shaped outside portion of your eye (and the part that hurts and drives you crazy when you get something in your eye). The cornea bends the light coming into it. This light-bending behavior is part of the process of focusing on what you're looking at.

2. **The iris controls the amount of light.** The iris, which is the colored part of your eye, closes or expands to allow more or less light to enter. When it's dark, the iris expands to allow more light to enter, but it contracts to limit an excessive amount of light.

3. **The pupil regulates the light.** When the iris expands or contacts, it exposes more or less of the pupil, the dark center of your eye that gathers the light. You can think of the iris as a part of the pupil's light-control system. A camera's aperture works a lot like the iris and pupil in your eye; it controls how much light is allowed to act on the camera's sensor.

4. **The lens focuses the light and passes it on.** The lens is the clear, inside part of your eye. It works with the cornea to focus the world around you, and then passes that focused image to your retina.

5. **The retina converts the image to electrical signals.** The retina is a layer of tissue at the back of the eye that contains *photoreceptors*, which are cells that interpret light and convert it into electrical signals.

6. **The optic nerve carries the signals to the brain.** The brain takes the signals from the optic nerve and interprets them so that you can understand the world around you. Working together, your eyes and brain can interpret what you see in as little as 13 milliseconds. (A millisecond is one-thousandth of a second).

Human Eye Anatomy

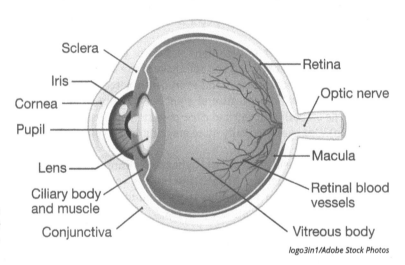

FIGURE 8-3:
Human eye
anatomy.

Exploring rods and cones

The part of the eyesight process that really affects drone pilots is the retina. As noted in the previous section, the retina is a layer of tissue at the back of the eye that contains cells called photoreceptors. There are two types of photoreceptor cells that allow you to see in the day and at night: cones and rods. If you take the Part 107 exam, you're very likely to see questions about these topics.

Considering cones

One of the two photoreceptor cells is called *cones* (see Figure 8-4). Cones are called cones because, well, they sort of look like cones! Cones enable you to see during the daytime, and they interpret color. Different cones in the retina are responsible for specific colors, and the cones work together to create color. For example, if you look at a yellow flower, red and green cones work together to create a yellow color signal that will be sent to the brain. The cones are capable of interpreting a massive number of colors and color shading.

TIP

Along with enabling color, the location of most of the cones (the macula, or *fovea* area) is what gives you central vision. The cones help you look directly at something in the center area of your vision and interpret both what you see and its color. Cones don't work well in the dark, however. That's where the second photoreceptor comes into play.

Understanding rods

Rods are the second type of photoreceptor cell, and once again, they're called rods because that's what they look like (also shown in Figure 8-4). Rods help you see in low-light conditions (technically called *scotopic vision*). However, rods do not interpret color; only cones do that (and as mentioned previously, cones don't work well in the dark). If you see a bowl of fruit in a dark kitchen at night, you can't distinguish the color of the fruit. Although the rods may help you distinguish the fruit in the bowl, you can't see any color.

Rods are located in the outer segments of the retina tissue, whereas cones are more centralized. This matters because in the dark, your central vision is poor; you see better using your peripheral vision instead. Now you know why you run into a piece of furniture at night in a dark house: It's hard to see what's directly in front of you, and you tend to lose spatial awareness when it's really dark.

REMEMBER

What does this issue mean for a drone pilot? If you're trying to fly and it's really dark, it's hard to see directly in front of you. Basically, the very front of your field of vision can be a sort of blind spot. Using sweeping motions with your eyes so that you engage more of your peripheral vision can help you see better. Keep that point in mind.

Photoreceptor cell

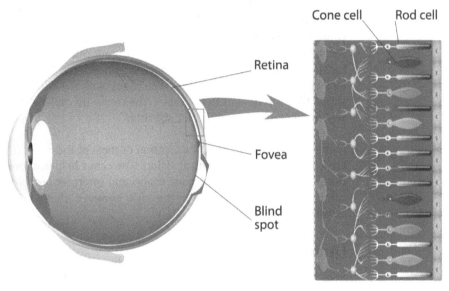

FIGURE 8-4:
Cone and rod
photoreceptor
cells.

designua/123RF

Adapting to dark conditions

For your eyes, adjusting from a bright environment to a dark one takes some time. When you go from a brightly lit room at night into a dark room, you don't see much at first. The details of the room soon start to take shape, albeit without color. If you stayed there for half an hour, you would see even better.

It takes your eyes a full 30 minutes to adapt to dark conditions. The cones don't do much and the rods have to take over, and that takes a bit of time. If you see any bright lights during that 30-minute window, the entire process has to start over.

REMEMBER

The FAA wants you to understand this principle. Your eyes need to fully adapt to night conditions before you fly. This means that you need to wait in the darkness for half an hour before taking off. If you need to get your drone ready to fly, and you turn on your phone's flashlight, that 30-minute window needs to start over.

For that reason, you might consider using a red light if you need a light source before you fly, such as a flashlight with a red filter over it. The red light doesn't interfere with your eyesight's night adaptation the way a white light does.

Understanding Night-Flight Visual Problems and Illusions

Although your eyes' rods enable you to see at night, your vision is not the same as in daytime. Visual problems and illusions can arise in the dark, and these issues can impact the safety of your nighttime flight missions. The potential problems covered in the following sections relate to light sources that may appear in your field of vision when you fly at night.

Phantom motion

Phantom motion occurs when a light source appears in the distance against a dark background. Think of a distant tower with lights against a night sky. You can see the lights, but those lights may appear to sway back and forth at a regular speed (oscillation). The lights may appear to be moving even though they aren't.

This kind of situation can create confusion, forcing you to differentiate between real motion and phantom motion so that you can maintain a stable and safe flight.

TIP

If you experience phantom motion, just slow down your flight speed or stop and hover for a moment until you can get your bearings. This phenomenon isn't preventable, but slowing down or stopping to assess your surroundings can help.

Fascination

Another potential nighttime visual problem, called *fascination*, occurs because our eyes and brains are naturally attracted to a light source. When you see a light in the distance, you can fixate on it even though you don't realize you're doing so. This fixation can cause you to lose situational awareness. Shockingly, fascination has been known to cause drone pilots to crash into an object because they became so fixated on the light source, they didn't realize where the drone was actually flying.

There's no fix but to realize this is a potential issue. You have to force yourself to ignore the light source and stay focused on the drone and the controller.

Autokinesis

Autokinesis is similar to phantom motion, but instead of swaying back and forth, the light source can appear to drift or move to a different location. This happens when a drone pilot becomes too fixed on the light source (this affect can work with fascination as well). If you believe a light source is moving, that will naturally affect your piloting decisions, which can lead to safety problems.

The solution to the problem is the same as for fascination: You must force yourself to stop looking at the light source and focus your attention on the drone and the controller.

Flicker vertigo

It's not unusual for some light sources to flicker at night, such as a tower light or some kind of beacon. This usually isn't a problem. However, if the flickering is fast, the flicker can cause dizziness, disorientation, and even motion sickness in some people. Naturally, you don't want to fly when you feel dizzy, disoriented, or sick.

WARNING

The key is to not focus on the light but on the drone. Still, some people have real problems with flickering lights like this, and if you're susceptible, land your drone quickly. You can always try the mission later. If you feel dizzy or sick, the odds of a piloting accident drastically increase.

Size-distance illusion

Objects that appear against a dark background often appear bigger than they actually are. This situation is simply an effect of your eyes being unable to distinguish the scale of an object or structure without other, comparative objects in your field of vision. Strangely, those structures or objects always appear bigger than they are instead of smaller.

Believing a structure to be bigger than it is may affect your piloting decisions and judgment calls. Try to scan your eyes in the general area to locate other items to visually grab onto as a way to help you understand the scale of the structure in question. That method may help. Otherwise, just be aware of this illusion as you fly.

Reversible perception illusion

The reversible perception illusion has a couple of important applications. First, know that when it's dark and you fly toward some kind of light source, that light source may start to appear larger than it actually is. The reverse is also true. If you're flying away from a light source, it may appear smaller than it actually is. In the former case, this illusion can make it difficult to estimate the actual distance to a light source because your perception isn't accurate.

This same difficulty of estimating distance comes into play with a moving light source. You often can't tell whether the light source is moving toward you or away from you. For example, say you're performing a piloting mission at night, and in the distance, you see a low-flying airplane. Is the airplane moving toward you or away from you? For safety reasons, you need to know because you may need to yield right of way.

Fortunately, airplanes have a light system in place to help you understand their direction. The following lights appear on airplanes (which is true for all crewed aircraft in the United States, except Alaska):

>> **Red:** The plane's left wing has a red light that is visible from the front or back of the plane.

>> **Green:** The plane's right wing has a green light that is visible from the front or back of the plane.

The plane also has a white light on its tail, but that doesn't particularly help with the reversible perception illusion.

Here's what you need to know. If you're flying your drone and you see an airplane in the distance, note where the red light is relative to you:

>> If the red light is to your *right,* the plane is flying *toward* you. ("Red to my right: Careful with my flight.")

>> If the red light is to your *left,* the plane is flying *away* from you. ("Red to my left: No need to be deft.")

Figure 8-5 can help clarify.

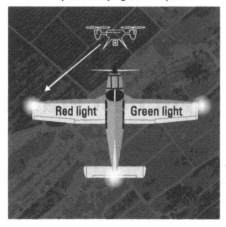

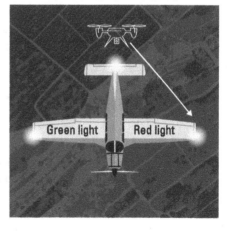

A **red light** on YOUR right means the plane is flying toward you.

A **red light** on YOUR left means the plane is flying away from you.

FIGURE 8-5: Airplane position lighting.

But what if you see only a green light or only a red light? This means that the other wing isn't visible because the plane is turning (the other wing is blocked from your field of view). In that case

>> If you see only a red light, the plane is turning to the left.

>> If you see only a green light, the plane is turning to the right.

The best I can do here is a little alliteration and a rhyme: "Red light: Left. Green light: Right."

Remember that you're looking at the plane, so when it's coming toward you, its left is on your right and vice versa.

TIP

You may think, "Can't I just look at the white taillight and know the plane's direction?" No —you really can't. In many cases, you won't be able to tell whether the taillight is behind the wing lights or in front of them (in terms of the plane's direction). This is a visual illusion because you can't see the rest of the plane to interpret where the taillight actually is. That's why the FAA uses red and green lights on the wings. Just remember that!

Developing Night-Flight Best Practices

So now you know a lot about flying at night, and it's time to put some final tips to work as you think about night-flight best practices. Luckily, you don't have too much to remember, but the few things you do need to remember are important, so read on!

Check out the area in the daytime

If you know you're going to fly at night in a particular area, it's a good idea to check out the area in the daytime. The daytime view will always be different than the nighttime view, and that's important.

REMEMBER

Say you have a drone piloting job that requires you to fly around some structures at night. If you can fly around those structures during the day and get familiar with everything, you'll find the night flight much easier because you'll have a daytime visual memory of the area to pull from. You will likely feel more confident and at ease with the night flight because the area won't be new to you.

Don't ignore pre- and post-flight checks

For some odd reason, when you plan a night-flight mission, it's easy to forget things you normally check on during day flights. However, the basic flight is the same. Always make sure that

>> The batteries and RC are charged and ready.

>> You've checked the weather forecast.

>> You've inspected the drone's propellers.

>> For night flights, you have anti-collision lighting installed on your drone and it is charged.

These are some of the items to remember. See Chapter 2 for more details about pre- and post-flight checks.

Don't forget LAANC approval if needed

I don't want to sound like a broken record, but if you need LAANC approval to fly in an area during the day, you also need that approval at night. Nothing changes with the U.S. airspace when the sun goes down. Also, night skies are full of other aircrafts, so make sure to use an app (such as AutoPylot or other FAA-approved app) to see whether you're planning to fly in controlled airspace and if you need LAANC approval.

Use off-center viewing and scanning

As noted earlier in this chapter, in the section "Exploring rods and cones," the rod cells in your retina are not located in the very center of your vision. Typically, at night, the very center of your vision is a sort of blind spot. For that reason, a couple of night-sight strategies can help you fly safely.

Use off-center viewing

When you use off-center viewing, you're aware that an object in the center of your field of vision won't be discernible. The strategy is to look 10 degrees above, below, or to the sides of the object you need to view. Doing so engages the rods in your eyes and helps you see better in the dark.

Use a scanning technique

A scanning technique helps you see better in the dark because you don't fixate on one area. Scanning techniques help you maintain better situational awareness (what's happening all around you as you fly).

There are a few different scanning techniques you can use, but here's an easy one. Mentally divide the area you're viewing into three or four segments (see Figure 8-6). Then, moving your eyes from left to right or vice versa, scan an area for two seconds or so before moving to the next area, all while keeping your drone in your field of vision. Just keep repeating this pattern as you fly.

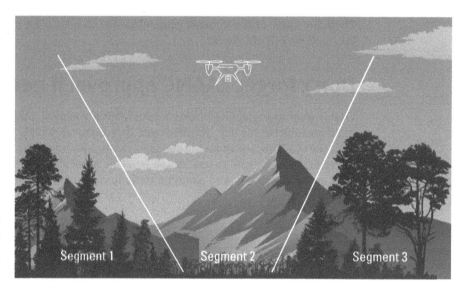

FIGURE 8-6:
A scanning technique can help you see better at night.

TIP

Scanning can help you avoid visual illusions and other potential problems that the night tends to bring with it. Also, night flights and scanning are easier if you have a visual observer with you who is doing the same thing. Together, you can keep everything you need to see in your field of vision, which will give you a safer flight.

Be physically and mentally ready

Finally, make sure you're physically and mentally prepared to fly at night. Night flights require even more concentration than daytime flights do, so you need to be at your best. Have you eaten lately? Did you drink enough water today (and no alcohol)? Did you get a good night's sleep last night? Are your eyes tired from the tasks of the day? Are any big stressors distracting your attention at the moment? These are a few of the questions you should ask yourself before you fly a drone at night.

3

Getting to Know the National Airspace System (NAS) and Sectional Charts

Chapter **9**

Checking Out the National Airspace System

The National Airspace System (NAS) in the United States governs and manages air traffic — everything from commercial airplanes to drones. Along with defining regulations and rules, the NAS manages procedures with airports, runways, and other kinds of aviation facilities. It's also responsible for aeronautical charts (more about that topic in Chapters 11 and 12) and all kinds of technical data and procedures that help keep the skies safe. The United States has around 45,000 flights a day with around 3 million people flying, covering the entire United States. It's a big system and extremely important!

The U.S. is made up of both *controlled airspace*, where air traffic control manages flights, and *uncontrolled airspace*, which air traffic control doesn't manage. As a drone pilot, you will interact with both controlled and uncontrolled airspace, depending on where you're flying, and the Federal Aviation Administration (FAA) expects you to understand the airspace system and the different classes of airspace, along with the rules governing each class. Keep in mind that a B4UFLY app will help you know which airspace you're flying in and the rules for that airspace. That's one reason the B4UFLY apps, which I talk about in Chapter 2, are so important to download and use.

In this chapter, you explore each of the airspace system classes and the rules that govern those classes. Also, before you begin, make sure you're familiar with MSL and AGL. If those concepts sound foreign to you, check out Chapter 6 before you start here!

TIP

Some of the figures in this chapter show you airspace on a sectional chart. Sectional charts are very important to understand if you're taking the Part 107 exam. I present close-up, cropped versions of the sectional chart; you can download the entire testing supplement at https://www.faa.gov/sites/faa.gov/files/training_testing/testing/supplements/sport_rec_private_akts.pdf.

In relevant figures, I also point out the figure number in the testing supplement so that you can check it out.

Class A Airspace

Class A airspace is probably the easiest airspace class to understand because it's simple, straightforward, and has nothing to do with drone piloting! Yet, you could see a Part 107 exam test question on it, and you need to know that it's there just for your understanding of the entire system.

REMEMBER

Class A airspace is a high-level airspace from 18,000 feet MSL to 60,000 feet MSL, as shown in Figure 9-1. Clearly, you're not flying your drone here, but this airspace is used by commercial airplanes cruising to a destination.

Class A airspace requires Air Traffic Control (ATC) permission (controlled airspace) to use and is used with instrument flight rules (IFR). In other words, commercial airplanes use instruments to fly here, not manual flying (thank goodness).

TIP

Keep in mind that Class A airspace is simply a cruising altitude for commercial airliners — it's higher than the other airspaces and doesn't even show up on aeronautical charts because it's above all the other airspace traffic and rules.

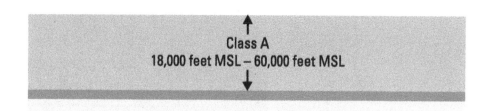

Class A
18,000 feet MSL – 60,000 feet MSL

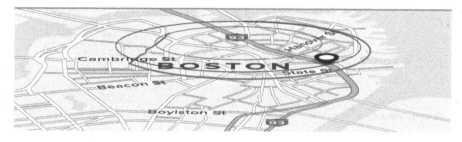

FIGURE 9-1:
Class A airspace,
18,000 feet
MSL to
60,000 feet MSL.

Class B Airspace

Class B airspace is used around the big airports in the United States. (When you see class "B," think "Big Airport.") There are 37 Class B airspaces in the country covering those big airports, such as JFK, LAS, DFW, and so on. Class B airspace is, of course, controlled airspace, so you need permission, often from Air Traffic Control (ATC), to fly a drone here.

REMEMBER

Class B airspace is custom designed for each airport and consists of different levels, or tiers within the airspace. In a drawing, this airspace is often compared to an upside-down wedding cake (see Figure 9-2).

Class B airspace sits on top of big airports, and you can think of each layer of the airspace as having a *floor* and a *ceiling*. The airspace begins at the floor and ends at the ceiling. So, at the lowest tier, or layer, the floor begins at that layer's ground level and extends up to its ceiling height. Then the next layer begins, and so on. The airspace is designed this way to accommodate planes entering the airspace for landing, as well as planes flying at certain altitudes for takeoff and eventually leaving the airspace.

REMEMBER

Typically, the first layer of Class B airspace starts at the ground level of an airport (what the FAA calls the *surface*) and extends upward to around 10,000 feet MSL. Each upper layer picks up where the layer under it ends and extends upward and outward from there.

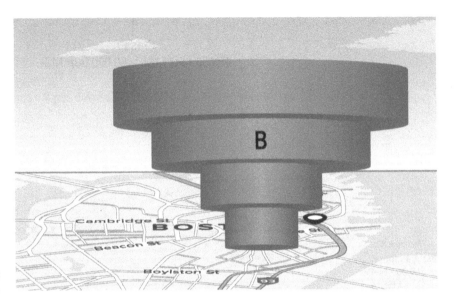

FIGURE 9-2:
Class B airspace.

Again, Class B is designed around an airport, so the floor and ceiling vary based on that airport. In order to know the floors and ceilings of each layer, you use an aeronautical chart. On the chart, Class B is shown with a blue solid line. As an example, San Francisco International Airport (SFO) is a class B airport, as shown in Figure 9-3. On the chart, you see solid blue lines around the airport showing the different layers of the airspace. These appear as rings circling SFO (and yes, sectional charts of Class B airports are very complex because so much is happening around those airports!).

The chart shows you the floor and ceiling in each layer. For example, the inner-most layer says 100/SFC. That means the ceiling is 10,000 feet. On charts, feet is represented in hundreds but it's actually thousands. The FAA does this to make the information shorter and easier to read. Just add two zeros to the number, which is 10,000 feet. SFC means "surface." So, the innermost layer of this Class B airspace is the surface (ground) to 10,000 MSL.

Other floors and ceilings in outer areas of the airspace appear on the chart. For example, start at the SFO runway on the chart and travel southwest into the next blue-lined layer. You see the number 100 over the number 15, which means that the ceiling of that layer is 10,000 feet MSL and the floor is 1,500 feet MSL.

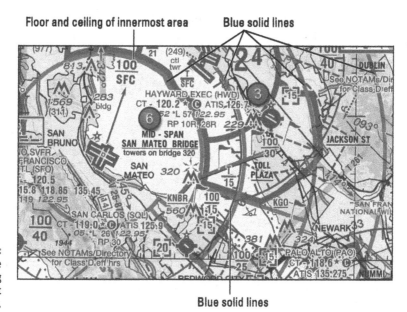

Floor and ceiling of innermost area

Blue solid lines

Blue solid lines

FIGURE 9-3:
Class B airspace
(testing
supplement
Figure 70).

TIP

Returning to drone piloting, because Class B airspace is in tiers expanding upward and outward (remember the upside-down wedding cake example), can you fly your drone in an outer area of the layers as long as you're not in the airspace? Yes, because in those cases, you're not yet in Class B airspace; you're in Class E or G (more about those in the Class E and Class G sections, later in this chapter).

Consider an example. Say you visit San Francisco, and you want to fly your drone near San Francisco International Airport (SFO). You've traveled from the airport to the west and now you're south of the toll plaza that you see on the chart, shown in Figure 9-4. You know that SFO's airspace is still above you. However, you're not in the first layer of airspace, which starts at the ground level. You're two layers away on the sectional chart. You see that the airspace layer is surrounded by a blue line (Class B), with a ceiling of 10,000 feet and a floor of 2,500 feet.

So, can you fly your drone here without permission? Yes, because you're not in Class B airspace until you reach the floor at 2,500 feet. Because your maximum legal altitude is 400 feet with your drone, you're good to go!

This is where understanding airspace floors and ceilings really comes into play. It's important that you can interpret this information using a chart, and remember, your B4UFLY app can help you with airspace restrictions as well. If I look on my app for where I want to take off southwest of SFO, I can confirm that I am in Class G airspace, shown in Figure 9-5, and I'm good to fly!

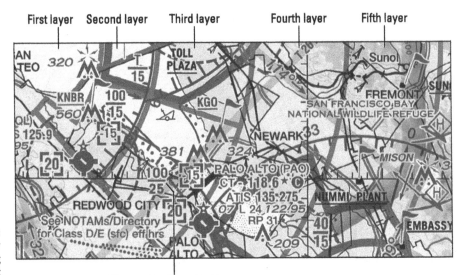

First layer   Second layer   Third layer   Fourth layer   Fifth layer

Floor and ceiling, third layer

FIGURE 9-4:
Class B outer
layer (testing
supplement
Figure 70).

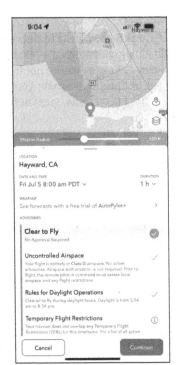

FIGURE 9-5:
Class G airspace
on a B4UFLY app.

Class C Airspace

Class C is similar to class B airspace — just smaller. Class C airspace is used at busy airports in cities, but not those really big airports that you see with class B airspace (think B = Big Airport, C = City Airport). Similarly to Class B, Class C airspace uses layers, but it typically doesn't have as many, as shown in Figure 9-6.

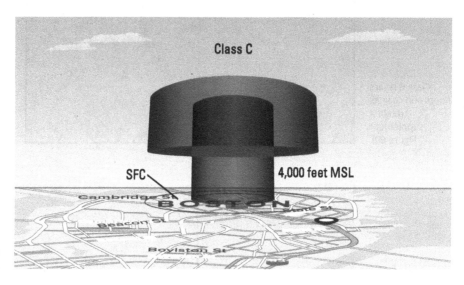

FIGURE 9-6:
Class C airspace.

Class C airspace is controlled by Air Traffic Control (ATC,) so you need permission to fly in Class C. Generally, Class C begins at the surface at an airport and extends up to 4,000 feet MSL. The outer-layer floors will begin at different levels and extend to around 4,000 feet MSL.

REMEMBER

On charts, Class C airspace appears as solid magenta lines, and you'll see the floor and ceiling notations written in the same manner as those for Class B. For example, Corpus Christi International Airport (CRP) is a Class C airport. Note on the chart the varying floors and ceilings for the different layers, as shown in Figure 9-7.

To make this practical, return to a B4UFLY app. Say you choose a location near CRP where you want to fly your drone. You check the app, which shows you where you are on the map, as shown in Figure 9-8. However, you see that you're in Class C airspace and need LAANC approval to fly here (see Chapter 7 for more information about LAANC approval). At the moment, on the chart, you're in the innermost area of the Class C airspace, which is controlled from the surface to 4,000 feet (refer to Figure 9-7), so approval is needed.

Solid magenta lines   Floor and ceiling in each layer

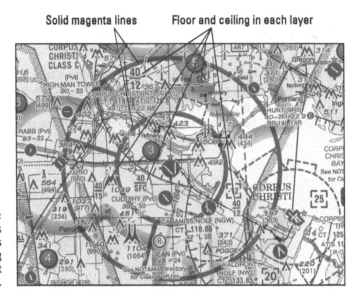

FIGURE 9-7:
Class C layers
floors and ceilings
(testing
supplement
Figure 69).

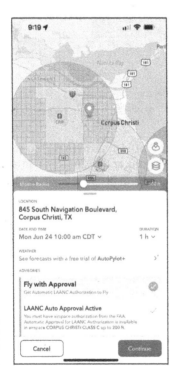

FIGURE 9-8:
LAANC approval
needed
to fly here.

However, what if you drive east for a few miles so that you can fly around Corpus Christi Bay? Refer to the chart in Figure 9-7. You're under the outer layer of Class C airspace, the floor of which is 1,200 feet. Because you're flying up to only 400 feet, you're flying under the floor of the airspace, which means you're actually in Class G — uncontrolled — airspace. The app confirms this fact, as shown in Figure 9-9, so you're good to fly here.

FIGURE 9-9: Class G, uncontrolled airspace under the Class B floor of 1,200 feet.

Class D Airspace

Class D airports surround airports that have a control tower, but they're not as big or as busy as Class C airports. They are represented as a single cone, so they don't have different layers of airspace like Class B and C. Typically, Class D airspace starts at the surface and extends up to 2,500 feet AGL, but this can vary. This airspace generally has a radius of 4–5 nautical miles, as shown in Figure 9-10. Class D airspace is controlled airspace, so you need approval to fly your drone in Class D.

Typical Class D Airspace Dimensions

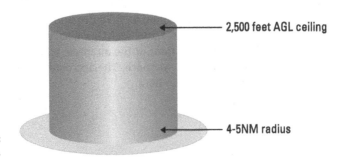

2,500 feet AGL ceiling

4-5NM radius

FIGURE 9-10:
Class D airspace.

Class D airspace starts at the surface and extends up to a certain level, represented on the chart as a number in a bracket (in MSL). On the chart, Class D airspace is represented with a dashed blue line. For example, in Figure 9-11, Minot International Airport (MOT) is a Class D airport with airspace from the surface to 42,000 feet MSL.

Dashed blue lines   Ceiling, 42,000 feet MSL

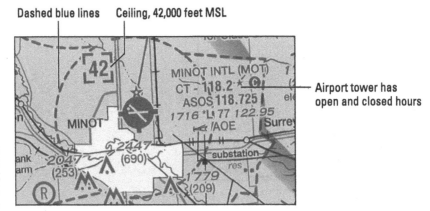

Airport tower has open and closed hours

FIGURE 9-11:
MOT is Class D airspace (testing supplement, Figure 21).

TIP

Be aware of a couple of important points about Class D airspace. First, you may see a ceiling number with a minus in front of it, such as −22. This means that the airspace goes up to 2,200 feet MSL but does not include that number. So, −22 means the airspace goes up to but not including 2,200 feet MSL. At 2,200 feet, the airspace converts to whatever airspace may be above it. This happens in cases with airspace directly above Class D. If another airspace starts at 2,200, the minus sign tells you that the Class D airspace goes all the way to 2,200, but not to it, because 2,200 is the floor of another airspace.

TIP

Also, some Class D airports may have a control tower that isn't open all of the time. For example, the tower at MOT isn't open during night hours. Airports that have towers with certain operational hours are also noted on the charts with a star beside the control tower's radio frequency, also shown in Figure 9-11 (and see Chapter 10 for more information).

WARNING

Why does knowing the control tower operating hours matter? When the control tower is not in operation, the Class D airspace converts to either Class E or Class G, which means you may not need permission to fly during those hours (Class G). However, note that airplanes still may take off and land even when the tower is closed. So even if you don't need permission to fly around an airport when the tower is closed, you still need to fly with great care and use situational awareness. Drone pilots who fly regularly around airports with closed control towers often invest in an ultra-high frequency (UHF) radio so that they can hear pilots announcing takeoff or landing intentions. See Chapter 10 for additional information about UHF radios.

Class E Airspace

Class E airspace is controlled airspace that is interspersed around Classes B, C, and D. There's a lot of Class E airspace in the United States, and it can be a bit confusing at first because there are a couple of different kinds. First, take a look at Figure 9-12, which shows a snapshot of an area mixed with several classes of airspace. Notice how Class E airspace sort of sits in between areas of other airspaces, specifically B, C, and D.

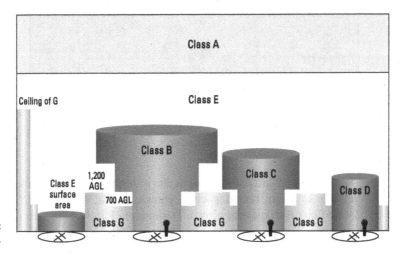

FIGURE 9-12:
Class E airspace.

Notice how, in Figure 9-12, Class E airspace fills in the gaps and Class A airspace sits above it. In most cases, Class E airspace extends up to meet Class A airspace, at just under 18,000 feet MSL. You also see Class G on the graphic; I explain Class G in an upcoming section.

Class E is often confusing because it consists of a few different types, as the following sections explain.

Starting from the ground up

Sometimes Class E airspace begins at ground level. This is especially true for small, regional airports that do not have a control tower. From the ground up, you're in controlled airspace around these airports.

REMEMBER

On a chart, Class E airspace from ground level appears as a dashed, magenta line. For example, Jamestown Regional Airport (JMS)) is a Class E airport and begins at the ground. As you can see on the chart, it is outlined with a dashed, magenta line, as shown in Figure 9-13.

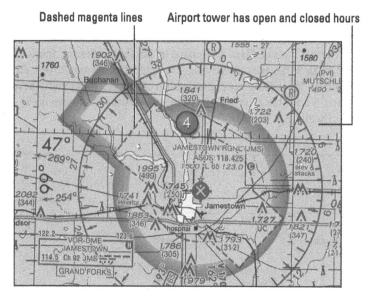

FIGURE 9-13:
Class E airport (testing supplement Figure 26).

Because Class E airspace begins at the ground in cases like this, you do need permission to fly in this airspace. If you check that on a B4UFLY app, you see that JMS's Class E airspace approval, as shown in Figure 9-14.

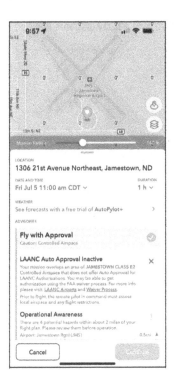

FIGURE 9-14:
Permission
needed to fly in
Class E airspace.

Starting from 700 feet AGL

The second type of Class E airspace begins at 700 feet AGL and can extend all the way up to but not including 18,000 feet MSL (the beginning of Class A airspace).

The 700-foot floor of Class E airspace is very common, and airspaces at this level appear around airports as a thick, faded magenta line, as shown in Figure 9-15. Some small airports have only this type of airspace, and some have a starting-from-ground-level Class E airspace surrounded by a 700-foot AGL Class E airspace in the outer area (refer to Figure 9-12).

For example, Garrison Municipal Airport (D05), shown in Figure 9-15, has only the 700-foot Class E airspace.

You don't need permission to fly in this airspace because this Class E airspace doesn't begin until 700 feet AGL (and your maximum legal ceiling with a drone is 400 feet AGL). You can check the airspace on a B4UFLY app. For example, when I fly in Garrison Municipal Airport's airspace, I see on my app that I can fly here without permission, as shown in Figure 9-16.

However, having permission doesn't mean the absence of all dangers. Planes still take off and land at these airports, even though they may be few and far between. Always use situational awareness when you fly in these airspaces.

700-foot AGL Class E Airspace shown with a fuzzy magenta line

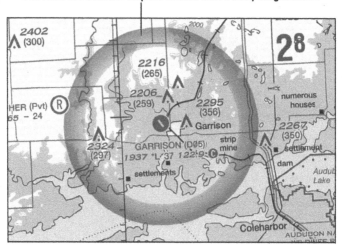

FIGURE 9-15:
700-foot AGL
Class E airspace
(testing
supplement
Figure 21).

FIGURE 9-16:
Permission isn't
needed to fly in
the 700-foot AGL
Class E airspace.

TIP

Do you notice that the airspace in Figure 9-16 says Class G? You may be thinking, "Wait — I thought Garrison Airport was Class E!" Remember that Class E here doesn't begin until 700 feet AGL. I'm flying under that floor, so I'm in Class G airspace, which the app confirms.

Starting from 1,200 feet AGL

Class E airspace beginning at 1,200 feet AGL is what is above most of the country, unless one of the other airspace classes overrides it. This airspace is not marked on charts and you don't need permission, but of course, you're not flying at 1,200 feet AGL anyway. For the most part, Class G airspace sits under this Class E airspace up to 1,200 feet AGL, at which point the airspace becomes Class E. That's all you really need to know, and of course Class G is where you fly a lot, so keep reading!

Class G Airspace

The final airspace classification is Class G, which typically starts at the ground and can continue up to but not including 1,200 feet AGL (refer to Figure 9-12). Remember, airspace in the U.S. is either controlled or uncontrolled. So far, this chapter has considered controlled airspace, but Class G is uncontrolled, and honestly, it makes up most of the country. You don't need permission to fly in Class G airspace unless there is something else in Class G that overrides that situation, such as a stadium or other special no-fly zone.

A sectional chart doesn't say anything about Class G because it's not called out. Just know that when you look on the sectional chart, if you're not in another airspace, you're in Class G, up to but not including 1,200 feet AGL (where it turns into Class E).

Types of Special-Use Airspace

Along with the NAS classes, the United States has a few special-use airspaces. This section takes a quick look at those and how they appear on sectional charts.

Warning

Warning areas on sectional charts are noted with a *W* followed by a number, and you'll see them in blue hashed lines. For example, in Figure 9-17, you see W–107A, W–107B, and so on. Warnings typically appear on charts as you leave the United States, such as when flying out over the ocean. You can still fly in a Warning area, but the warnings are given to alert you to potential dangers, such as military training and other aerial activities.

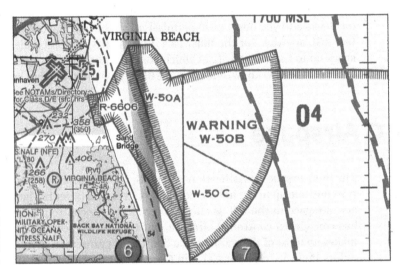

FIGURE 9-17: Warning area (testing supplement Figure 20).

Restricted

Restricted airspace is just as the name sounds. It's restricted, so you can't fly in this space without permission. These are often military training areas that may contain artillery fire or other dangers. Sometimes these areas are not active, however, and you can get permission to fly if you need it.

On sectional charts, Restricted areas are outlined with blue hash lines and contain an *R* followed by a number, such as R-5401 and R-5402, as shown in Figure 9-18.

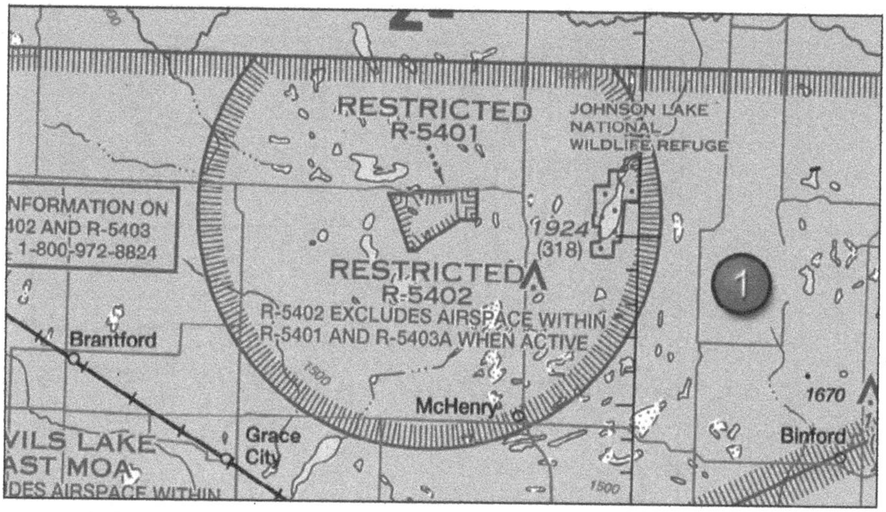
RESTRICTED
R-5401
JOHNSON LAKE
NATIONAL
WILDLIFE REFUGE
NFORMATION ON
402 AND R-5403
1-800-972-8824
1924
(318)
RESTRICTED
R-5402
R-5402 EXCLUDES AIRSPACE WITHIN
R-5401 AND R-5403A WHEN ACTIVE
Brantford
McHenry
1670
Binford
VILS LAKE
AST MOA
Grace
City
1

FIGURE 9-18: Restricted area (testing supplement Figure 26).

Prohibited

As you might guess, you can't fly in a Prohibited area — ever. These special-use areas are designed for national security and surround specific places that may endanger certain governmental officials. For example, Camp David, the presidential retreat in Maryland, is a Prohibited area.

On a chart, a Prohibited area appears with a *P* followed by a number, such as P-40, as shown in Figure 9-19. Restricted areas appear with blue dashed lines, and the interior of the Prohibited area appears in white. (This map does not appear in the testing supplement.)

Alert

An Alert area is an area that may have a high volume of aircraft or other unusual aerial activity. You can fly your drone in an Alert area without special permission, but just be aware that it may contain hazards.

An Alert area appears with an *A* followed by a number, and sometimes the sectional chart will tell you why the area is an Alert area. For example, in Figure 9-20, A-231 designates a student jet training area, and the chart tells you so. Alert areas are outlined with magenta-colored hash marks.

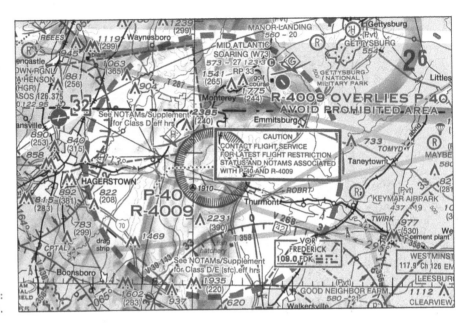

FIGURE 9-19:
Prohibited area.

FIGURE 9-20:
Alert area (testing
supplement
Figure 75).

Military Operations Area (MOA)

A Military Operations Area (MOA) is a military training area. In an MOA, you may experience low-flying aircraft. You don't need permission to fly in an MOA, but you do need to use situational awareness.

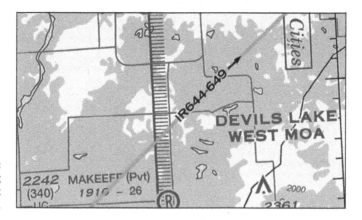

FIGURE 9-21:
MOA (testing
supplement
Figure 21).

On a sectional chart, MOAs are shown with magenta hash marks, and you'll see the name of the MOA inside of those lines. For example, in Figure 9-21 you see the Hog B MOA, and you even see some warnings to pilots about rapidly rising terrain. Again, you can fly here without permission; just use good situational awareness.

DON'T FORGET A B4UFLY APP!

Airspace rules and regulations can get confusing, and if you're taking the Part 107 exam, you'll need to know the details of them, so study carefully. Practically, as you're drone piloting, you may need to check out a chart from time to time in complex flying locations, but as you plan to fly, don't forget to use a B4UFLY app. These apps are extremely helpful because they can tell you about airspaces and restrictions. They will also tell you if you're in a prohibited, restricted, or warning area. These apps save a lot of time and help you fly in a legal and safe manner. Check out Chapter 2 for details. Don't forget the app!

Chapter **10**

Getting to Know Confusing Acronyms and Airport Radio Communications

L ike almost all industries, the aviation industry is full of acronyms and principles represented by various words. Just working through the meaning of different aviation verbiage can be challenging! Yet, some of these acronyms have particularly important meanings because they can affect where and how you can fly. They work alongside the airspace concepts you find in Chapter 9 and in sectional charts in Chapters 11 and 12, so it's a good idea to know what these words mean and how to access the information they represent.

Along with that, the Federal Aviation Administration (FAA) expects drone pilots to know some things about airport radio frequencies as well.

TIP

If you're taking the Part 107 exam, you can expect to see some questions about these different concepts, so you need to know what these acronyms mean and how to understand them. But don't worry —I tell you only what you really need to know! In this chapter, you check out confusing acronyms, some basics of sectional chart notations, and the airport communication concepts you need to understand.

Understanding Sectional Charts

It's a dreaded term in the droning world: *sectional chart*. If you've been reading this book from the beginning, you know I've mentioned sectional charts a couple of times, as needed, but for this chapter and the next two chapters, you spend much more time understanding and interpreting sectional charts.

People dread sectional charts because they can be hard to read and confusing at first. Think of them as aerial maps that tell you everything you need to know about flying in a particular area. The problem is, you have to think both horizontally and vertically; after all, you're not flying on a flat piece of paper.

In a nutshell, sectional charts are important for all aircraft pilots because they tell you what is going on in a particular airspace. You just have to learn to read and understand them.

So, to begin finding out about sectional charts, download the FAA's testing supplement called *Airman Knowledge Testing Supplement for Sport Pilot, Recreational Pilot, Remote Pilot, and Private Pilot*. The document number is FAA-CT-8080-2H, and you can download it at https://www.faa.gov/training_testing/testing/supplements/media/sport_rec_private_akts.pdf.

This document is the testing supplement the FAA uses on the Part 107 exam (and from here on, I just call it the *testing supplement*), so I use it in this chapter for examples as you discover more about what's available in the charts and in the guide overall. As you explore the document, I tell you what chart you're seeing in this book, and you can check out the full-color version of that chart in the supplement.

TIP

Note that the charts give you airspace information, but they often include a bunch of other information as well. It's helpful to think of a chart as providing *layers* of information. Also, be sure to train yourself to think vertically as you look at the chart because what's happening in the air is what a chart represents.

The next section explores some specific acronyms, notations, and airport radio frequencies because this information may appear on various charts as well.

Exploring Common Acronyms and Notations

As mentioned previously, there are several acronyms related to charts that you need to know, and honestly, plenty more that you don't have to worry much about as a drone pilot. You can find a list of legends and definitions in the testing supplement toward the beginning of the document. However, note that there are nineteen different pages of legends! That's a lot of information to learn, but you don't need to know all of it because the training supplement applies not only to drone piloting but also to private piloting as well. So, don't worry about memorizing these pages or even reading all of them! I help you know what's important to take a look at.

NOTAM

A *NOTAM* is a Notice to Air Mission, which is effectively a notice about a particular airspace area that may be temporary or that notes something unknown at the moment. Basically, sectional charts give you static information about an airspace, but the NOTAM is designed to give you new information as it becomes available that could impact a drone flight.

For example, if a large public event or perhaps a natural disaster of some kind is occurring in an area, the NOTAM will give you up-to-date information about flying in that area. So you won't find NOTAMs on a sectional chart, but you should check for them before you fly in any area.

You can check for NOTAMS at the FAA at `https://notams.aim.faa.gov/notamSearch`. You can use different search parameters here, but an easy method is to search for the airport you'll be flying around. For example, in Figure 10-1, I'm searching the DFW airport area.

FIGURE 10-1:
A NOTAM search.

For DFW, the search results give me a bunch of NOTAMs — honestly, an overwhelming number because it's such a big airport. Of course, many of these NOTAMs don't apply to drone pilots because they discuss runways and other items. You can use the filter option at the top to choose different types of NOTAMS. For example, if you just want to see airspace NOTAMS, you can choose that filter option, as shown in Figure 10-2.

FIGURE 10-2:
NOTAM search
results and
filtering.

REMEMBER

After you find a NOTAM you want to view, just click it to review the details. Typically, the NOTAM provides location information and any other information you need to know. Remember, NOTAMS aren't specific to drone pilots, so you have to review these to see what, if any, apply to you in a particular area. Just keep in mind that they can be helpful if there are recent developments, damage from storms, temporary obstructions (such as construction cranes and other temporary structures), or particular events that will bring a lot of people to the area.

TFR

A TFR is a Temporary Flight Restriction, which is a kind of NOTAM that restricts flights completely or below a certain height ceiling. A TFR may stop you from flying in an area for a period of time, so it's an important consideration for drone pilots.

TIP

TFRs are issued in cases of natural disasters, special events where flights are not allowed, or the movement of certain government personnel, such as the president. Typically, any time the president is in an area of the country, you'll see a TFR as long as the president is there. As you can imagine, TFRs are issued for safety reasons, and, in some cases, to prevent too much air traffic in a particular area that may be unsafe.

The most common TFRs are issued based on the movements of the president and vice president, so on any given day there are TFRs in various areas of the country. Often, those TFRs are limited, and they usually do not have a big impact on the airspace system as a whole. Just know that TFRs are issued nearly every day. Also, as another example, big events such as the Super Bowl come with TFRs as a means of reducing airspace traffic.

As previously mentioned, flight restrictions are in place around any stadium that holds 30,000 or more people. These are considered *standing TFRs* because they're always in effect when an event is occurring at these stadiums. (See Chapter 7 for more information about flying over people.)

WARNING

Violating a TFR can result in fines and even the suspension of your Part 107 certificate, and it's the pilot's responsibility to be aware of TFRs and abide by them. So you need to check for them before you fly in an area where a TFR could be in effect. Usually, unless I'm flying in a very rural area, I check it first!

The good news is that it's easy to check for TFRs. Just visit https://tfr.faa.gov/tfr2/list.html to see a list. You can filter this list by an urban center or state; also, at the top of the window you can switch to a map view and zoom in to a particular area, as shown in Figure 10-3.

If you see a TFR in an area where you want to fly, just click it to view the details. For example, in Figure 10-4, this TFR was issued due to space flights at Cape Canaveral. The TFR tells you the start and end date, location, and the restriction. In this case, the TFR prohibits any kind of flights from the ground to 18,000 feet MSL. So, basically, no flying here!

TIP

It's important to note that TFRs often restrict flights from the ground up to a certain height. For example, commercial airplanes may be able to fly over a TFR because they're flying above the height ceiling. Drone pilots are almost always affected because we fly only to 400 feet. So always play it safe and check for TFRs because they may be in effect where you want to fly.

MTR

An *MTR* is a Military Training Route, which is considered a low-altitude, high-speed training route for military planes. And when I say high speed, I'm talking about as high as 300 miles per hour in some cases!

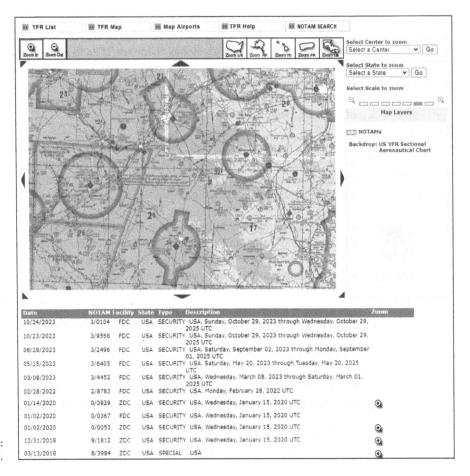

Date	NOTAM	Facility	State	Type	Description	Zoom
10/24/2023	3/0104	FDC	USA	SECURITY	USA, Sunday, October 29, 2023 through Wednesday, October 29, 2025 UTC	
10/23/2023	3/9568	FDC	USA	SECURITY	USA, Sunday, October 29, 2023 through Wednesday, October 29, 2025 UTC	
08/28/2023	3/2496	FDC	USA	SECURITY	USA, Saturday, September 02, 2023 through Monday, September 01, 2025 UTC	
05/15/2023	3/6405	FDC	USA	SECURITY	USA, Saturday, May 20, 2023 through Tuesday, May 20, 2025 UTC	
03/08/2023	3/4452	FDC	USA	SECURITY	USA, Wednesday, March 08, 2023 through Saturday, March 01, 2025 UTC	
02/28/2022	2/8783	FDC	USA	SECURITY	USA, Monday, February 28, 2022 UTC	
01/14/2020	0/3929	ZDC	USA	SECURITY	USA, Wednesday, January 15, 2020 UTC	⊕
01/02/2020	0/0367	FDC	USA	SECURITY	USA, Wednesday, January 15, 2020 UTC	
01/02/2020	0/0053	ZDC	USA	SECURITY	USA, Wednesday, January 15, 2020 UTC	⊕
12/31/2019	9/1812	ZDC	USA	SECURITY	USA, Wednesday, January 15, 2020 UTC	⊕
03/13/2018	8/3984	ZDC	USA	SPECIAL	USA	⊕

FIGURE 10-3:
TFR map view.

Because MTRs are low-altitude routes, they're important considerations for drone pilots. There are two kinds of MTRs, and you'll see notations about them on sectional charts:

>> **IFR:** Flights above 1,500 feet are considered *instrument flight rules (IFR),* which means the pilot is using instruments for the flight.

>> **VFR:** Flights below 1,500 feet are considered *visual flight rules (VFR),* meaning the pilot uses their eyes to see where they are flying.

MTRs will appear on sectional charts and are noted as IR (IFR) and VR (VFR), with a number next to them. For example, in Figure 10-5, you see IR678. Routes with four numbers mean the route is at 1,500 feet or below, and routes with three numbers are above 1,500 feet. So IR678 is an instrument flight rules route above 1,500 feet.

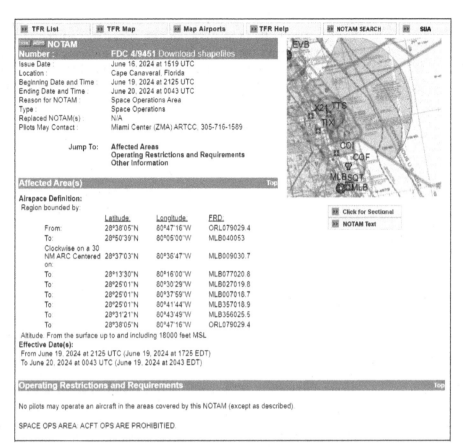

FIGURE 10-4:
TFR details.

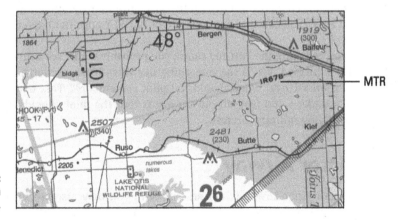

FIGURE 10-5:
MTR IR route on a
sectional chart.

You can legally fly in an MTR, and IR routes above 1,500 feet aren't a concern because you can't fly that high with a drone anyway, unless you're doing some special tower inspection. The VR routes can travel below 1,500 feet, so they *are* your concern because you don't know how low those flights are going to go. Figure 10-6 shows a VR route.

MTR

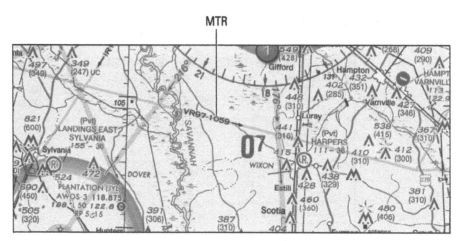

FIGURE 10-6:
MTR VR route on
a sectional chart.

As you think about MTRs, TFRs, and NOTAMs in general, you'll be glad to know that many B4UFLY apps give you this information. For example, in Figure 10-7, the AutoPylot app shows information about MOAs (Military Operations Area where MTRs are likely to appear), and any TFRs and other NOTAMs. Being provided with this quick information and related web links can be very helpful. So once again, always check your B4UFLY app to make sure that flying in any given area is safe and legal. You can find out more about B4UFLY apps in Chapter 2.

TIP

If you're taking the Part 107 exam, you probably won't see a direct question about MTRs. Instead, you'll see something like this: "On a sectional chart, you see the notation VR1437. What should the pilot remember?" This notation indicates an MTR under 1,500 feet, so the answer is that the pilot should be on the alert for low-flying military aircraft.

Victor Airway

Though *Victor Airway* is not an acronym, you may also see Victor Airways on a sectional chart. They are depicted by light-blue lines with a label, such as V21. Victor Airways are important for drone pilots because they are low-altitude, civilian-aircraft airways from 1,200 feet AGL up to 18,000 MSL, and are classified as Class E airspace.

FIGURE 10-7:
Information may
be provided on a
B4UFLY app.

REMEMBER

Basically, the FAA wants you to know about Victor Airways because you can encounter low-altitude planes, so just remember that if you're flying in an area with a Victor Airway notation, on a sectional chart (see Figure 10-8).

Victor Airways

FIGURE 10-8:
Victor Airways on
a sectional chart.

VFR Checkpoint

A *VFR* is a *Visual Flight Record* checkpoint (also called a waypoint) that appears on sectional charts as a magenta-colored flag. VFR checkpoints are simply notations about larger intersections, rivers, lakes, and other geographical features. You can fly here, but you may see more air traffic in these areas, so the sectional charts point them out, as shown in Figure 10-9.

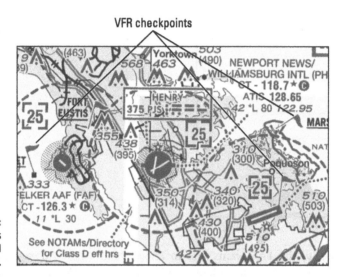

VFR checkpoints

FIGURE 10-9:
VFR checkpoints
on a sectional
chart.

Deciphering Airport Radio Communications

You may be surprised to know that the FAA wants drone pilots to know a thing or two about accessing airport radio communications, so much so that the Part 107 exam will likely ask you at least one question about it. Part 107 pilots are not required to have a VHF radio to hear those communications., but you may need to fly in controlled airspace near an airport — especially a nontowered airport, where it's a good idea to know what pilots are saying in terms of takeoffs and landings. In that case, having a VHF radio and understanding how to access those communications can help you keep your drone out of the way of other aircraft around the airport.

TIP

Fortunately, the sectional chart for a particular airport tells you about the radio communications for that airport; you just have to know how to read it. (I provide more information about towered and nontowered airports in Chapter 13.).

UNICOM and CTAF

Big airports have control towers that communicate with pilots. You can monitor these communications, but you should keep yourself out of it. You need training and even additional certifications to participate in such communications. Non-towered airports use frequencies that provide automated information about the airport, as well as frequencies that enable pilots to self-identify and report their intentions (such as landing). There are two kinds of frequencies (and the frequency number varies from airport to airport) as follows:

>> **UNICOM (Universal Communications):** Information on the UNICOM frequency is broadcast from a station and often provides information about the airport, conditions, and anything else pilots need to know. Pilots can use UNICOM to get information about runways and other important information regarding takeoff and landing. You can think of UNICOM as an information service.

>> **CTAF (Common Traffic Advisory Frequency):** CTAF is a frequency that pilots use to talk to each other and announce intentions, such as current location and landing. At nontowered airports, pilots use CTAF to talk to each other so that they can coordinate position, altitude, and takeoff and landing intentions. Basically, they talk to each other to avoid collisions and accidents.

With a UHF radio, you can monitor both UNICOM and CTAF around an airport, which can help you know what's going on and what pilots are saying to each other. I talk about UNICOM and CTAF a bit more in Chapter 13, but I want to define the terms here as well as mention that the information for each airport is available in both a chart supplement and directly on the sectional chart for the airport.

All that said, here's an example of a CTAF frequency on a sectional chart. Say that you want to monitor CFAF communications at Cooperstown Municipal Airport. You get the sectional chart for Cooperstown and see a C with a circle around it, along with a radio frequency in front of that C. You're seeing the CFAF frequency, 122.9, as shown in Figure 10-10.

Along with CFAT, you can access UNICOM to gain general information about the airport, weather, and so on. Sometimes the CFAF and UNICOM frequencies are the same, but not always. The UNICOM frequency is usually 122.8, but you can check the chart supplement to be sure (see Chapter 13 for more about the chart supplement). Also, the UNICOM frequency may be listed after the airport's runway length on the sectional chart.

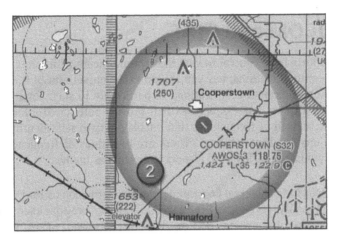

FIGURE 10-10:
CTAF frequency
at Cooperstown.

CT and ATIS

For bigger airports, you'll also see radio frequency information for the control tower (CT) and Automatic Terminal Information Service (ATIS). First, the control tower frequency will say CT, followed by the frequency. For example, in Figure 10-11, the control tower frequency for the Savannah airport is 119.1. Also, notice the star icon following 119.1, which indicates that the control tower is in operation part time. When the control tower is not in operation, you use CTAF, explained in the previous section. Depending on the sectional chart, sometimes the control tower information can be hard to see because so much other information is crowded around it. It's there — you just have to look closely!

Control Tower Frequency

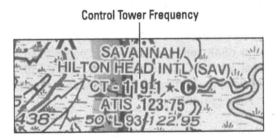

FIGURE 10-11:
Control tower
frequency
at Savannah.

ATIS automated information is continually broadcast at busier airports. It provides weather information, runway information, and other details that crewed aircraft pilots need to know. ATIS is also listed on the sectional chart. You'll see the word "ATIS" followed by a frequency number, as shown at the Norfolk International airport in Figure 10-12.

ATIS frequency

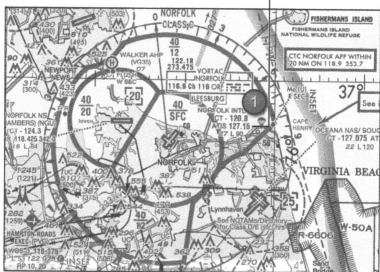

FIGURE 10-12:
The ATIS
frequency at
Norfolk
International
airport.

RADIO COMMUNICATIONS PHRASEOLOGY

If you monitor airport radio communications, you may come away from that experience saying, "I have no idea what language the pilots are speaking." That's because airport communications follow certain patterns and phraseology. That phraseology is a completely new thing that you can learn, but as a drone pilot, you're not required to do so. Still, if you plan on flying around airports and monitoring CTAF or even CT communications, you may want to learn some basics just so that you can follow along. An FAA internet page breaks down all the phraseology. There's a lot and it can be confusing, but it's also interesting. Take a look at https://www.faa.gov/air_traffic/publications/atpubs/aim_html/chap4_section_2.html.

Chapter **11**

Discovering How to Read Sectional Charts

S ectional charts give you specific information about a particular area in the United States. You can find all kinds of information including airspace, radio frequencies, topography information, and much more. Basically, the chart is designed to help you know what's going on in the air and potentially around the ground in that area.

The FAA expects drone pilots to be able to read and use sectional charts, and if you're taking the Part 107 exam, you can expect a few questions that require you to use a sectional chart and interpret information from that chart. Often, sectional charts are the most dreaded aspect of the Part 107 exam, but if you're a little bit of a nerd like me, you may just find them interesting and sort of cool to be able to read and understand.

In this chapter, you find out how to interpret the markings and notations on a sectional chart, with emphasis on what you need to know as a drone pilot. I tell you about latitude and longitude, elevation notations, and dealing with obstructions, and you'll get a bit of sectional chart interpretation practice to wrap things up. When you can understand the various pieces of the chart, you have a much better shot at dissecting all the data that's crammed together on it.

TIP

One note: It would be best to read Chapters 9 and 10 before this one so that you're familiar with airspace and a few specifics that appear on sectional charts. This chapter's text assumes that you're familiar with the concepts in those chapters.

Understanding Latitude and Longitude

Latitude and longitude are coordinates on a map that help all kinds of pilots navigate the planet. They also appear on sectional charts, so you need to be able to find a location on a sectional chart using these coordinates. Happily, reading coordinates isn't too difficult after you get the hang of it. Latitude and longitude coordinates are expressed in degrees, minutes, and seconds. Alternatively, GPS expresses coordinates in a digital latitude and longitude format. (I explain the digital aspect in "Understanding GPS digital coordinates," later in this chapter.)

First, some definitions will help:

>> **Latitude:** *Latitude* refers to lines that run parallel to the equator. Latitude tells you how far north or south of the equator something is located. The equator is considered zero latitude. On sectional charts, above the equator is the Northern Hemisphere and below it is the Southern Hemisphere. Degrees above the equator are expressed in positive numbers, such as 20, 40, 60, whereas degrees below the equator are expressed in negative numbers, such as –20, –40, –60 (see Figure 11-1). As a point of reference, the United States falls between 25 degrees N and 49 degrees N (N means "parallel North." The N tells you that the 25 degrees noted is 25 degrees above the Earth's equatorial plane).

>> **Longitude:** *Longitude* refers to the measurement that runs horizontally from the North Pole to the South Pole. The prime meridian is Greenwich, England, which is considered zero longitude. Whereas latitude shows you locations North and South of the equator, longitude represents locations east and west of the prime meridian. Degrees of longitude are expressed from 0 to 180 degrees in both the east and west (see Figure 11-1). The United States is located between 67 degrees and 120 degrees west longitude.

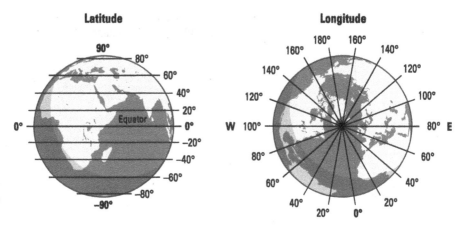

Reading locations on a sectional chart

See the preceding section for explanations of latitude and longitude. This section shows you how to read locations on a sectional chart.

You can download the sectional charts that you'll see on the Part 107 exam from https://www.faa.gov/training_testing/testing/supplements/media/sport_rec_private_akts.pdf. I also note in each figure caption the testing supplement map you see so that you can view it in full color in the PDF download. Also, you can check out the sectional chart for any airport in the United States at https://vfrmap.com/.

When you look at a sectional chart for an airport, the sectional chart notes (somewhere on the chart) the airport area in terms of latitude and longitude. Ignoring everything else on the chart for a moment, notice the parallel and horizontal lines showing latitude and longitude to find degree notations somewhere on the chart. For example, Figure 11-2 shows 48 degrees latitude intersecting with 101 degrees longitude.

Of course, if all you need to know is the particular location on the chart where the chart tells you the coordinates, you're home free. But in real life (and the exam), you'll need to use the given coordinates as a base and then figure out other coordinates yourself. So you need to know how to locate coordinates and determine the coordinates where a particular point is located.

To do that, it's important to understand how the lines are displayed, and that comes down to degrees. A full degree, such as moving from 100 degrees to 101 degrees longitude, is 60 "minutes." Each minute is marked on the longitude or latitude line, so you can count each minute for specific accuracy. Notice that every tenth mark is bigger, which can help you keep on track, sort of like using a ruler or measuring tape.

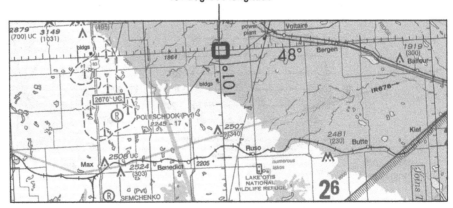

48 degrees latitude
101 degrees longitude

FIGURE 11-2:
Latitude and
longitude
intersection (see
testing supple-
ment, Figure 21).

For example, in Figure 11-3, you can see the 98-degree and 99-degree longitude lines. If you count the marks shown on the chart between 98 and 99, you'll find that there are 60 minutes, which equals a full degree. This same concept is true when you consider latitude. You're moving north or south, but you'll find that the latitude lines are 60 minutes apart, creating one full degree. Minutes are expressed with a minute mark, which is a single quotation mark. So, 26 minutes is written as 26'.

Also, you'll see a line at 30 minutes between two full degrees. For example, in Figure 11-3, you see a line between 98 and 99 degrees. This means that you're at the halfway point between the two degrees, or 98° 30'.

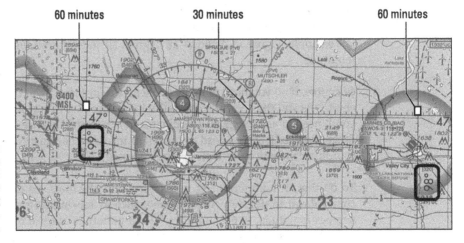

60 minutes   30 minutes   60 minutes

FIGURE 11-3:
60 minutes
equals one full
degree
(see testing
supplement,
Figure 26).

REMEMBER

It's important to keep in mind that if you're traveling west longitude (*away* from the prime meridian), the number increases by one degree every 60 minutes. If you're traveling east, *toward* the prime meridian, the number decreases by one degree (because you're traveling toward the prime meridian, which is 0). Also, for latitude, if you're traveling north away from the equator (zero degrees), the number increases by one degree every 60 minutes, but it decreases by one degree every 60 minutes if you're traveling south.

To bring all these concepts together, take a look at Figure 11-4. Note the star I've placed on the sectional chart. Where is this star in terms of latitude and longitude?

Star location

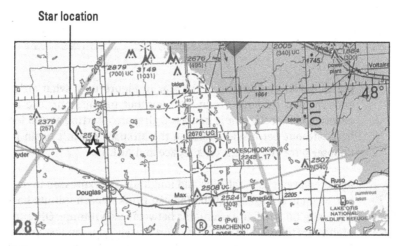

FIGURE 11-4:
Where is the location of the star? (testing supplement, Figure 21).

You figure out the location of the star by taking it one step at a time. To find the longitude of the star, note the longitude line for 101°. Because the star is to the west (left), you count up from 101. But you also see the 30-minute line, so you can jump over to 101° 30' and start from there. If you keep counting the tick marks on the line, you can see that the star is about two marks past the 30' mark, so the longitude of the start is 101° 32'.

Next, look at the latitude. On the chart, you see the line for 48°. However, the star is below the 48° line, so you need to start counting backward (moving south). Remember that each degree is separated by 60 minutes, so if you start at 60 and count down, you see that the star sits next to 55 minutes. So, the latitude is 47° 55'.

Try another example. Take a look at Figure 11-5. Locate the star on the chart and see if you can calculate the latitude and longitude of the location.

Did you find it? The star's location is about 46° 55' N latitude and 98° 8' W longitude.

Fuzzy, magenta lines   Star location

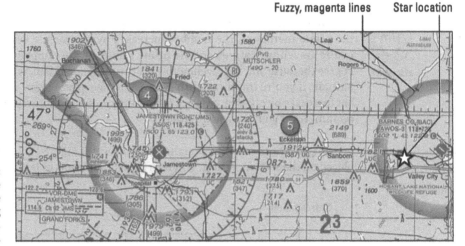

FIGURE 11-5:
Where is the
location of the
star? (testing
supplement,
Figure 26).

Bonus question: Looking at the star in Figure 11-5, what airspace is the star in, and can you fly your drone here without permission? The answer to the first question is that the star is in Class E airspace, starting at 700 feet AGL. You can fly here without permission because the airspace doesn't begin until 700 feet. (See Chapter 9 for the details on airspace and AGL.)

TIP

As noted previously, latitude and longitude are expressed in degrees and minutes, but seconds are used as well. You can think of each minute hash mark on a chart as having 60 smaller hash marks between each minute. Of course, that would be a nightmare to read and count, and fortunately, you can just ignore this information. For passing the Part 107 exam, degrees and minutes are enough to get you close without having to be exact. Just know that if you see a notation such as 48° 22' 46", the 46" means seconds, but again, you can ignore it for all practical purposes.

Understanding GPS digital coordinates

As mentioned previously, latitude and longitude are displayed as degrees and minutes on charts. However, Global Positioning System (GPS) uses a digital equivalent of degrees and minutes. So if you see a GPS coordinate, you need to convert that digital equivalent back to degrees and minutes in order to find the location on a sectional chart. A GPS coordinate will appear in decimal form, such as 42.61 N Latitude. To find 42.61 N Latitude on a chart, you need to convert the 61 to minutes so that you can find it. Fortunately, the 42 means 42°, so there's nothing you need to do. However, the minutes (.61) need to be converted (because if you assume that .61 is 61 feet, your chart location will be wrong). Fortunately, the conversion is easy to do. Just remember that there are 60 minutes in one degree, and multiply the decimal .61 by 60 minutes, which equals 37 (rounding up).

So, 42.61 in digital is 42° 37' in degrees and minutes.

Touching on Isogonic Lines

Sectional charts also display *isogonic lines*, which can be confusing. The good news is that you probably don't need to know anything about them as a drone pilot, unless you're planning a mission with a compass (which I doubt you are). However, you can get them mixed up with other lines on a chart, and the Part 107 exam may ask you a question about them as well.

Isogonic lines show magnetic variations between true North and magnetic North. This means that if you were using a compass and you saw an isogonic line on a sectional chart, you would need to adjust the compass heading.

Isogonic lines appear on sectional charts as dashed magenta lines with a degree variation noted on it. For example, in Figure 11-6, you see the isogonic line with 14° 30' E noted.

Again, you don't need to do anything with isogonic lines for the sake of drone flights, but I point them out to avoid any confusion.

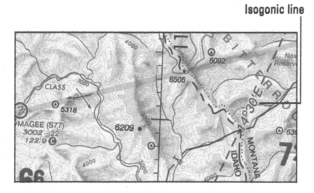

FIGURE 11-6: Isogonic lines (testing supplement, Figure 22).

Understanding Elevation

A final issue to consider with sectional charts concerns how charts convey elevations. Specifically, charts can give you information about both natural and human-made obstructions, and of course, obstructions are serious considerations for drone piloting.

Natural elevation maximums

Of course, the natural terrain across the United States isn't flat, and elevation matters, especially for airplane pilots. In other words, how high do you have to be flying to clear all of the natural terrain, such as mountains? Fortunately, the sectional charts help you know that information because the charts are in color. The different colors in the landscape represent maximum Mean Sea Level (MSL) ranges, and when appropriate, a sectional chart provides a legend on the page.

In Figure 11-7, the color chart on the right gives you the MSL elevation of different color areas on the chart. This particular area falls into the 7,000-foot MSL range, meaning that to clear all natural obstacles, you would need to fly above 7,000 feet MSL.

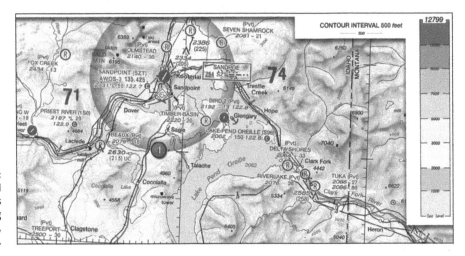

FIGURE 11-7: Natural elevations (testing supplement, Figure 22).

TIP

Keep in mind that this chart refers only to natural elevations, not human-made obstacles. To make sure you clear all natural and human-made obstacles, read on.

Maximum Elevation Figures (MEF)

Along with natural obstructions, there can also be ones constructed by people, such as buildings and towers. Pilots need to know the maximum elevation of those obstructions as well, so each quadrangle on the sectional chart displays a maximum elevation figure (MEF).

Figure 11-8 shows a big 6 and a little 6 next to each other in the middle of the quadrangle. The big number means thousands of feet, and the smaller number means hundreds of feet. The MEF for this quadrangle is therefore 6,600 feet MSL. Pilots need to fly at or above that elevation to avoid all obstacles.

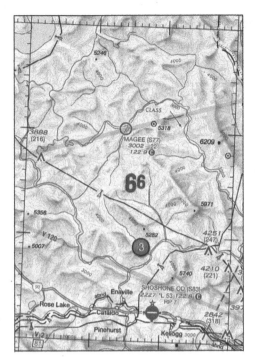

FIGURE 11-8:
MEF notation
(testing
supplement,
Figure 22).

Here's how the FAA figures the MEF, in case you're curious: They take the tallest obstruction (whether natural or human made), round that number up to the nearest 100, and then add another 100 feet just to be safe.

That's all you need to know about MEF, and of course, as a drone pilot, you're not flying anywhere near this high. But the FAA expects you to know what these numbers mean.

Towers and other obstructions

Sectional charts may point out potential obstacles, such as towers and other tall structures. The Obstructions section of the chart supplement gives you information and explains the icons used, as shown in Figure 11-9. There are icons for towers of certain heights and lighted towers, as well as other notes. *Note:* Elevation notations appear in MSL as a top number; under that number, the AGL appears in parentheses.

Following is an example of a chart displaying obstructions. Figure 11-10 shows a chart with three towers: two single towers and one group of towers (see the legend in Figure 11-9). Beside (or it could be above, or below . . . geez!) the icon are the MSL and AGL of each tower or tower group. In this example, the group of towers is 2,988 MSL and 288 AGL.

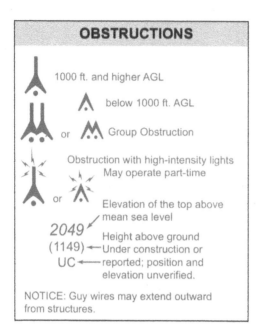

FIGURE 11-9:
Obstruction icons
(testing
supplement,
Legend 1).

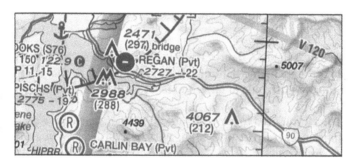

FIGURE 11-10:
Tower
obstructions
(testing
supplement,
Figure 22).

TIP

As you can see, obstruction height notations are rather straightforward, but sometimes the Part 107 exam will try to mix up MSL and AGL just to confuse you on questions. Just remember that the top number is MSL and the bottom number, in parentheses, is AGL.

Pop Quiz! Bringing Everything Together

Chapters 9 and 10 and this chapter show you different parts of the sectional charts. But in real life and on the Part 107 exam, you need to consider the entire chart and identify various aspects of it. In Chapter 12, I provide various practice scenarios and questions, but before I wrap up this chapter, I bring all the concepts from this and the previous two chapters together to help you gauge your current understanding. Figure 11-11 shows part of a sectional chart with the colored lines identified for those reading the print book. Look at the chart carefully and see whether you can answer the upcoming questions. (Don't worry — you can then read on for the answers!) Although the chart is provided in Figure 11-11, a color version is easier to interpret. You'll find these questions easier if you download the testing supplement at `https://www.faa.gov/training_testing/testing/supplements/media/sport_rec_private_akts.pdf`. And, with the downloaded chart, you can zoom in or out as needed to see small areas.

REMEMBER

If you take the Part 107 exam, you'll get a printed booklet (in color) of the charts. You won't have access to a digital version, so no zooming in or out on the test. That's why you may want to take a magnifying glass with you. Read more about this issue and discover more test tips in Chapter 20.

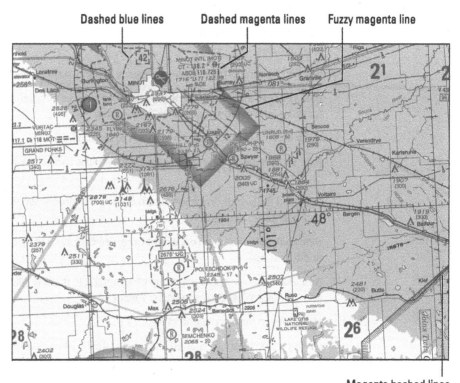

Dashed blue lines   Dashed magenta lines   Fuzzy magenta line

Magenta hashed lines

FIGURE 11-11:
Sectional chart (testing supplement, Figure 21).

Questions:

1. What is the approximate latitude and longitude of the Minot airport runway?

Hint: Refer to "Understanding Longitude and Latitude," earlier in this chapter.

2. You've been asked to fly near the Minot airport runway for a drone piloting job. What airspace are you in, and can you fly here without permission?

Hint: You need to look at the colored lines surrounding the airport to determine the airspace. See Chapter 9 to find out more about airspace.

3. What is the ceiling of the airspace immediately around Minot airport?

Hint: The ceiling notation is on the chart; you just have to find it! See Chapter 9 for more information about airspace floors and ceilings.

4. Just southwest of Minot airport is a lighted tower. What is the height of the tower in AGL?

Hint: The lighted tower icon's height is presented in MSL and AGL. Don't get the two confused!

5. What is the control tower radio frequency at Minot airport?

Hint: You find the frequency beside the control tower notation. See Chapter 10 for more information about radio frequencies.

6. Is the Minot airport control tower open all the time?

Hint: You're looking for a specific notation with the control tower information on the chart.

7. You need to fly your drone just south of Hill (PVT) airport. What airspace are you in, and can you fly here?

Hint: Look for colored airspace lines here. See Chapter 9 to find out more about airspace.

8. Near Poleschook Airport is a group of windmills. In the box with the elevation, you see "UC." What do these letters mean?

Hint: Refer to the icon legend in Figure 11-9.

9. On the chart is a small, shaded line labeled IR678. What does this line mean?

Hint: If you have trouble finding it, look toward the lower-right corner of the chart. The line is south of the towns of Bergen and Belfour.

10. On the bottom right of the chart, you see a series of hashed, magenta lines. What does this line mean?

Hint: This is a type of special-use airspace. See Chapter 9 for more information.

Answers:

1. The approximate latitude and longitude of the Minot runway is 48° 15′ N Latitude and 101° 15′ W Longitude. Find the degree notations near the airport, and then count the minutes to get to the airport's location on the chart.

2. The immediate area around Minot is Class D airspace (which you can tell by the dashed blue line), and you must have permission to fly here. Chapter 9 explains airspace and flight restrictions.

3. The ceiling of the Class D airspace is 4,200 feet MSL and is noted on the chart as 42 in a bracket.

4. The lighted tower is 1,031 feet AGL (and 3,149 feet MSL). Refer to Figure 11-9 for the icon legend (which doesn't appear in this practice figure). MSL is the bigger number, and your eye naturally notices the bigger number first. However, the question asks for AGL (the smaller number in parentheses beneath MSL). The Part 107 exam will mix MSL and AGL to try to confuse you, so be careful!

5. The control tower frequency is 118.2. Notice the CT under the airport name? That means "control tower."

6. No, the control is not open all of the time. A star next to the control tower frequency denotes that the control tower is operational in a part-time manner.

7. The fuzzy magenta line tells you that you're in Class E airspace that begins at 700 feet AGL (as explained in Chapter 9). Because you're flying under the floor of this Class E airspace, you don't need permission to fly here.

8. UC means "under construction." Refer to the icon legend in Figure 11-9.

9. IR678 is an MTR (Military Training Route). IR means the pilots are flying by "instrument rules." See Chapter 10 for more information about MTRs.

10. Magenta hash-marked areas denote a MOA (Military Operations Area). These are training areas for pilots. You can fly your drone here, but use caution because there could be low-flying military aircraft. See Chapter 9 for details.

How did you do? If you got confused or couldn't remember something, don't worry. It takes time and practice to interpret sectional charts. With that thought in mind, I've provided an entire chapter just for additional practice and to note some confusing and tricky questions the FAA likes to ask on the Part 107 exam. Turn to the next chapter to continue!

Chapter **12**

Practically Interpreting Sectional Charts

The FAA expects drone pilots to understand and be able to interpret sectional charts, and it includes some questions about these charts on its Part 107 exam. When you take the exam, you are provided with a testing supplement (which is a booklet of sectional charts and other information), but you can download this same testing supplement beforehand so that you can practice. (Sorry, but you can't take a digital version of the supplement into the testing center. You can use only the printed version they provide.) Download the testing supplement at https://www.faa.gov/training_testing/testing/supplements/media/sport_rec_private_akts.pdf.

In this chapter, I bring everything together that you explore in this part of the book. I provide some questions and scenarios that give you yet more practice (you get a little in Chapter 11) interpreting a sectional chart and making drone piloting decisions. (Note that this chapter won't make much sense to you if you haven't read Chapters 9–11 yet.)

With that said, if you're not planning to take the Part 107 exam soon, should you bother with this chapter? Yes. You should still get familiar with sectional chart interpretation so that you understand how different airspaces, obstacles, and other flight issues can impact drone piloting. Also, although sectional charts are often frustrating, you may that find they're fun to explore!

TIP

Although this book isn't a Part 107 test prep book, I offer questions and scenarios that are similar to the Part 107 exam so that you can get familiar with the FAA expectations of drone pilots. So take the following sections as a challenge to explore, think, and have fun! I present the sectional charts as figures, but you'll find these exercises much easier if you download and use the color charts from the website given in the first paragraph of this chapter. The caption for each figure in this chapter directs you to the chart figure you should look at.

Also, before you get started, check out Legend 1 in the testing supplement you downloaded. The legend shows you numerous icons used on sectional charts, which may help you answer some of the questions in this chapter correctly.

Zoning In on Questions about Airspace and Airports

Naturally, one of the most important pieces of information a sectional chart gives you is about airspace. The Part 107 exam uses sectional charts to test your knowledge of airspace and restrictions that may apply.

To help prepare you for the exam, this section shows you some charts, presents a series of questions to consider with each chart, and then follows up with the answers to those questions. I tell you the relevant chart figure number in each heading; you can look at the chart here or download the testing supplement (see the first paragraph in this chapter for the link) and view the chart in color.

Testing supplement: Figure 20

This section gets you started with a few smaller airports and explores the airspace around them. Take a moment to look over the chart in Figure 12-1 (Figure 20 in the testing supplement).

Questions to try

Here's a chance to try some practice questions. You can check out the answers provided after them.

1. To the West of Elizabeth City Regional Airport (ECG), you need to fly a drone mission near Center Hill. What airspace is this, and can you fly here without permission?

2. What is the control tower frequency at ECG?

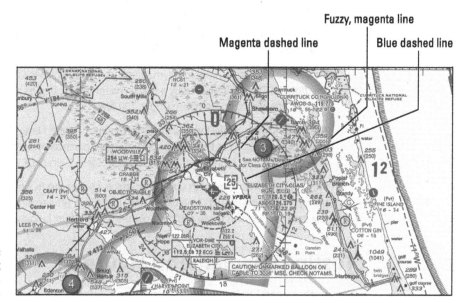

Magenta dashed line

Fuzzy, magenta line

Blue dashed line

3. Just south of ECG, you need to fly a drone mission at Weeksville. What airspace is this, and can you fly here without permission?

4. To the southwest of ECG, you need to fly a drone mission at New Hope. What airspace is this, and can you fly here without permission?

5. There is a group of towers slightly northwest of ECG. The tower group height is 334 feet AGL. You've been hired to fly your drone here to take inspection photos of the towers. Can you fly here without permission??

6. To the east of ECG is Pine Island (Pvt) Airport (*Pvt* means "private"). You need to fly directly next to this airport for a photo shoot. Can you fly here without permission??

Answers to those pesky questions

Find out how you did on the previous set of questions!

1. Center Hill is located in Class G airspace, and you don't need permission to fly here. Of course, the maximum height you can fly is 400 feet. Remember that everything on a sectional chart that is not marked as another airspace is Class G.

2. Locate the airport information on the chart about ECG. You see CT for "control tower" followed by a frequency, which is 120.5.

3. Weeksville is inside the blue, dashed line around ECG. This line indicates Class D airspace, which begins at the surface. You need permission to fly here.

4. New Hope is located in the fuzzy, magenta airspace, which is Class E airspace. The floor of Class E, indicated by a fuzzy, magenta line, begins at 700 feet. You can fly here without permission under 700 feet.

5. The towers are partially located inside a magenta, dashed line. This is Class E airspace that begins at the surface, so you need permission to fly here. (See Chapter 9 for the distinctions among types of Class E airspace.)

6. Pine Island (Pvt) is located in Class G airspace, so you can fly here without permission. Even though Pine Island (Pvt) is a small, private airport, you should use caution because there could be planes taking off or landing

Testing supplement: Figure 21

Try another chart! Before you dive into the questions, spend a moment looking at Figure 12-2, or better yet, look over the color chart in Figure 21 in the testing supplement.

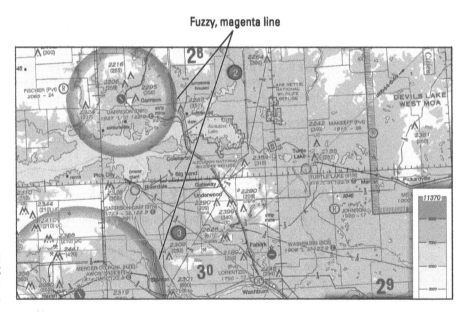

Fuzzy, magenta line

FIGURE 12-2: Testing supplement, Figure 21.

Questions

Here are your questions for the testing supplement's Figure 21. Maybe you'll ace them!

1. Just north of Garrison airport is a tower that is 259 feet AGL. You need to fly to the top of the tower for an inspection, but not above it. Can you fly here without permission?

2. You've been hired to take a series of drone photos spanning the entire length of Audubon Lake. Do you need permission to take these photos?

3. You've been asked to fly a mission one mile from Mercer County Regional (HZE). Can you fly this mission without permission?

4. You've been asked to inspect a tower in the Devils Lake West MOA. Can you fly this mission without permission?

Answers

How did you do on the preceding questions? Here are their answers:

1. Yes, the tower is located in Class E airspace with a floor of 700 feet AGL. You do not need permission to fly under the floor.

2. Audubon Lake is mostly in Class G airspace, but a segment of Class E airspace overlaps it. You can fly in this Class E airspace without permission because the floor doesn't begin until 700 feet.

3. Yes. HZE is in Class E airspace with a floor of 700 feet. You can fly here because you're under the floor. However, anytime you fly close to an airport, you should use caution because there may be airport traffic near you.

4. Yes. An MOA (Military Operations Area) is a military training area. You can fly here without permission, but you should exercise caution because there could be low-flying military aircraft.

Testing supplement: Figure 25

REMEMBER

With the chart in the testing supplement's Figure 25, things get more complicated! The DFW area airspace, shown in Figure 12-3, is very complex (and you'll probably see it on the exam). Be sure to download the testing supplement and look at the chart in color.

Solid blue lines

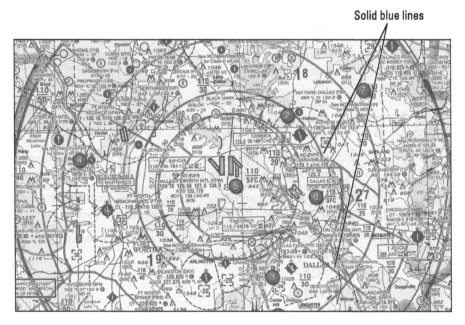

FIGURE 12-3:
Testing
supplement,
Figure 25.

Questions

When you're ready, dive into this next round of questions, for the chart in Figure 25 of the testing supplement:

1. What is the ATIS frequency at Dallas Ft Worth International Airport (DFW)?

2. What is the MEF at Air Park-Dallas (F69)?

3. Locate Dallas Executive Airport (RBD) on the chart (it's southeast of the DFW airport). What airspace is RBD in, and what are the floor and ceiling?

4. Find Arlington (GKY) airport on the chart. Near the airport is "–20" in brackets. What does this notation mean?

5. What is the floor of the airspace surrounding DFW?

6. To the far west of DFW is Eagle Mountain Lake. If you fly your drone here, what airspace are you in, and what is the floor?

Answers

I'll never leave you hanging. Here are the answers to the Figure 25 questions:

1. Locate the information about DFW on the chart. You see the ATIS frequency displayed as 123.775. The DFW area is a complex airspace area, and this overcrowded chart is frequently included on the Part 107 exam. Remember: The information is there; you just have to find it!

2. MEF means Maximum Elevation Figure (see Chapter 11). In Chart 25, the MEF is shown near the airport with a large 1 and a smaller 8, which indicates 1,800 feet. So the maximum elevation a plane needs to fly to clear all natural and man-made obstacles in this area is 1,800 feet MSL.

3. RBD is surrounded by a solid blue line, so RBD is in Class B airspace. There is notation near RBD that shows the number 110 over the number 30. This notation means that the ceiling of this Class B airspace is 110,000 feet MSL and the floor is 3,000 feet MSL.

4. When you see a negative number in brackets, this means that the ceiling of airspace extends up to but not including the number. In this case, –20 means that the airspace extends up to but not including 2,000 feet MSL.

5. If you look at DFW on the map, you see that the Class B airspace shows the number 110 over "SFC." This notation means that the ceiling of the airspace is 110,000 feet MSL and the floor is the surface (SFC — meaning at the ground; see Chapter 9).

6. You're still in Class B airspace here, as indicated by the blue lines around Eagle Mountain Lake. The notation of the number 110 over 60 tells you that the floor of this Class B airspace is 6,000 feet MSL.

Focusing on Elevations and Locations

Along with questions about airspace on the Part 107 exam, you're likely to see some questions about different elevation figures as well as latitude and longitude locations. These questions aren't as difficult as airspace questions, so this section looks at one chart from the testing supplement and provides you with just a few questions (and their answers, of course).

Testing supplement: Figure 26

Take a look at Figure 12-4 (Figure 26 in the testing supplement) and locate the latitude and longitude lines before you tackle the questions. Doing so will help you get mentally organized with the chart before you begin answering questions.

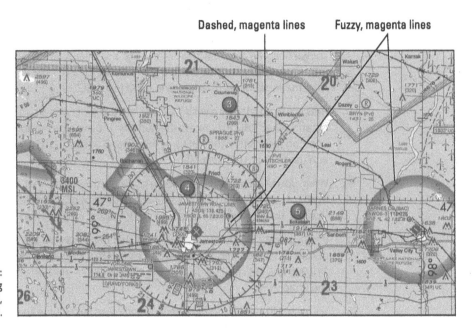

Dashed, magenta lines   Fuzzy, magenta lines

FIGURE 12-4:
Testing
supplement,
Figure 26.

Questions

Here's your practice quiz for Figure 26:

1. What is the highest elevation you can fly at Jamestown Regional Airport (JMS) without permission?

2. What is the highest you can fly at Barnes County Airport (BAC)?

3. What is the approximate latitude and longitude of Bryn (Pvt) airport?

4. What is the highest you can fly at Sprague (Pvt) airport without permission?

Answers

If you have trouble with the questions, hang in there! It gets easier with practice, and eventually you'll have all this down pat. Here are your answers to the questions for Figure 26:

1. Elevation and flight questions like this are basically airspace questions. In this case, JMS is surrounded by a dashed, magenta line. This line signifies Class E airspace that begins at the surface (which means at the ground). You can't fly here without permission at all. (See Chapter 9 for details.)

2. The answer is 400 feet. This is a trick question, so think carefully! BAC is Class E airspace with a floor of 700 feet, but as a drone pilot, your maximum height is 400 feet unless you're inspecting some tower or structure. In those cases, you can fly 400 feet above the tower. Otherwise, your height limit is always 400 feet.

3. On the chart, locate the nearest latitude and longitude notations, which are 47° and 98°. Then locate Bryn and count the minute marks to get the latitude and longitude, which are 47° 11′ North Latitude and 98° 9′ West Longitude.

4. Again, this is a kind of trick question. Sprague is in Class G airspace, so you don't have any airspace restrictions, but as a drone pilot, your maximum height is 400 feet.

Dodging Towers and Obstructions

You are likely to see questions about towers and other obstructions on the Part 107 exam, and sometimes the exam will try to trick you by mixing up MSL and AGL on the answer options. Just keep things straight on that front and you'll do fine. This section offers some examples of these types of questions.

Testing supplement: Figure 24

Spend a moment looking over Figure 12-5 (Figure 24 in the testing supplement). Before you tackle the questions, note the different airspaces you see on the chart.

TIP

It helps to always start with airspace when you look at a sectional chart. Getting familiar with the different airspace classes on a chart can help you answer questions correctly, especially as you start looking at the details on the chart (like towers and obstructions). It's easy to forget about the airspace and possible restrictions when you look at the details. Don't forget about airspace!

Questions

After acclimating your eyes a bit to Figure 24 on the testing supplement, try your hand at these questions:

1. Just to the east of Majors (GVT) airport is a tower that is 705 feet MSL. How high can you fly above this tower without permission?

2. There is a tower directly south of Card (Pvt) airport that is 1102 feet MSL. What is the maximum height you can fly above this tower in MSL?

Dashed, blue lines   Fuzzy, magenta lines

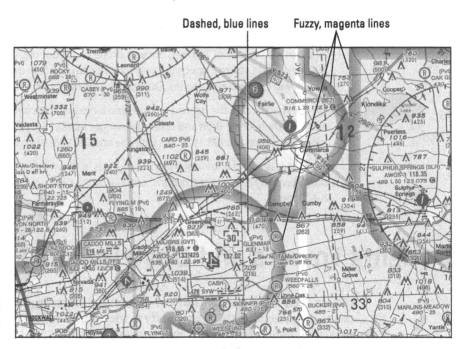

FIGURE 12-5:
Testing
supplement,
Figure 24.

3. Locate the lighted tower just south of Westminster. What is the AGL height of this tower?

4. What is the highest you can fly above the tower near Wolfe City?

Answers

There were some tricky ones on this round! Give yourself a pat on the back for any you got right (and another one just for trying).

1. You can't fly here without permission. This question will get you looking at the tower, but if you're not careful, you can forget about the airspace. The tower is in Class E airspace from the surface (dashed blue line). You can't fly here without permission.

2. This question is meant to be confusing because you tend to think in AGL with drone flights. The principle is the same as with MSL, however. You can fly 400 feet above a tower's height for inspection. In this case, the tower is 1,102 feet MSL, so just add 400 to that number and you get 1,502 feet MSL.

3. Towers and other obstructions typically provide the MSL (top number) and the AGL in parentheses under it. In this case, the tower is 700 feet AGL.

4. The tower is 309 feet AGL. You can fly 400 feet above the tower, for a total height of 709 feet AGL.

Testing supplement: Figure 25

You're back to the dreaded DFW sectional chart again, shown in Figure 12-6 (and Figure 25 in the testing supplement). Don't get overwhelmed. All the information you need is there, so take your time to find it!

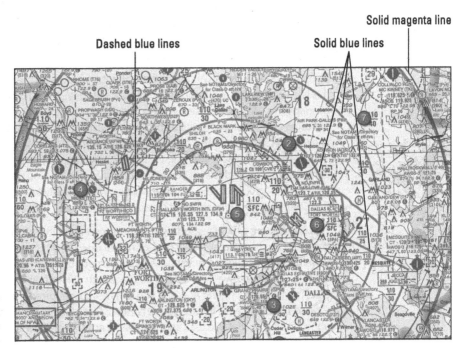

FIGURE 12-6:
Testing
supplement,
Figure 25.

Questions

I have just three practice questions for you this time:

1. At Ft. Worth Alliance (AFW) is a lighted tower cluster next to the runway. What airspace is this tower in, and can you fly here without permission?

2. You've been hired to photograph a business with your drone in Seagoville, southeast of DFW. Can you fly and photograph this property without permission?

3. Locate the lighted tower cluster just north of Allen (northeast of DFW airport). What airspace is the tower in, and what is the maximum height you can fly to inspect the tower?

Answers

In case you found the questions for the testing supplement's Figure 25 challenging, I offer some extra help in the following answers:

1. First locate AFW, but make sure you're looking at the actual airport runway instead of the text notations about AFW. If you find the runway, you'll see that the tower cluster is beside it. The airport and lighted tower are inside a dashed blue line, which means this is Class D airspace. You need permission to fly here. (See Chapter 9 for airspace restrictions.)

2. If you find Seagoville, you'll see a solid magenta line outside the area. Tracing this line reveals that it runs around a big area of DFW. A solid blue line also overlaps this same area. Don't get too bogged down here. A solid blue line indicates Class B airspace, and a solid magenta line means Class C. As best as you (and I) can tell from the chart, the Class B airspace here doesn't start until 4000 feet, so technically you're in Class G. You don't need permission to fly here. As a practical matter, if you were to fly this mission, you would check a B4UFLY app to confirm the airspace where you're actually taking the photos. See Chapter 2 for more information about B4UFLY

3. This is the same kind of inspection question you encounter previously in this chapter. (The FAA seems to like these kinds of questions.) In a nutshell, this question is an inspection question with an airspace question thrown in to try to confuse you. The tower is in Class B airspace with a floor beginning at 4,000 feet. The tower is 813 feet AGL and you can fly 400 feet directly above it, for a total of 1,213 feet AGL. In case you're curious, Class B airspace technically sits above the tower at 4,000 feet, so the tower is actually in Class G airspace.

Staying Alert to Special-Use Areas

Finally, you may see just a few questions about special airspace. These questions aren't complicated, but you need to remember what the acronyms mean (see Chapter 10). Here are a few sample questions.

Testing supplement: Figure 75

Spend a moment looking over the chart in Figure 12-7 (Figure 75 in the testing supplement). Also, you may need to access Legend 1 in the testing supplement so that you know what the icons mean.

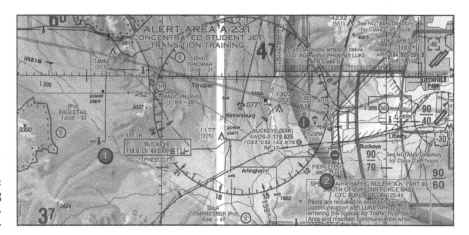

FIGURE 12-7:
Testing
supplement,
Figure 75.

Questions

Only two questions this time! But your ability to spot these items on the chart will help keep you and others safe.

1. What are two potential hazards near Buckeye (BXK) airport?

2. What does the line labeled V16 near BXK mean?

Answers

Now you can see how you did:

1. If you look around BXK, you can find a parachute icon. (It's slightly northeast, close to the airport.) This icon warns pilots that parachute training occurs in the area. Second, if you look above the airport, you see a notice that says, "Alert Area A-231." The warning tells you the area is a student jet training area. You can fly here, but you should use great situational awareness.

2. The letter *V* followed by a number means Victor Airway. This is a military training route that could have low-flying traffic. (See Chapter 10 for more about Victor Airways.)

Testing supplement: Figure 69

Take a look at Figure 12-8 (Figure 69 in the testing supplement). Be sure to look at the entire chart before you drill down to specific details. In other words, look at the big picture first!

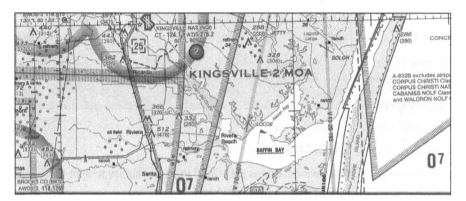

FIGURE 12-8:
Testing
supplement,
Figure 69.

Questions

If you've gone through all the quizzes in this chapter so far, you're down to your final one:

1. A realtor has asked you to take some photos of Cayo Del Grullo, a body of water. Can you legally fly here?

2. You want to fly around Baffin Bay. Are there any airspace restrictions?

Answers

Remember, there's no shame in not acing these quizzes yet. Understanding sectional charts takes time. See how you did on this one:

1. This area is in Class G airspace, but it sits inside an MOA (explained in Chapter 9). You can legally fly here, but you should be aware that there could be low-flying military airplane traffic.

2. No. Baffin Bay is in Class G airspace.

QUESTIONS ABOUT SECTIONAL CHARTS THAT HAVE NOTHING TO DO WITH CHARTS

The FAA does ask some questions on the Part 107 exam about charts, except that some of those questions don't have anything to do with charts. For example, you may see a question that tells you to refer to a certain chart in the testing supplement. Then, the question may ask you what the minimum visibility a drone pilot must have in a certain location, so you spend your time looking for something about visibility on the chart. But guess what? Charts don't tell you anything about visibility. The minimum visibility you should have is three nautical miles, regardless of where you're flying. The chart is just a distracter and a time waster.

Or, a question may direct you to a certain tower on a chart that is in Class G airspace, and then ask you how high above the tower you may fly for an inspection. The answer is 400 feet. You can fly 400 feet above the tower's height for an inspection in Class G airspace regardless of the tower's height. You don't have to look at the tower on the chart. These kinds of questions just waste your time, so here's a tip: Always read the entire question before you bother to find the chart in the test supplement. In some cases, you don't need it!

IN THIS CHAPTER

» **Exploring chart supplements**

» **Checking out towered and nontowered airports**

» **Understanding airport traffic**

» **Considering the right of way**

Chapter **13**

Flying Drones around Airports

When I was first studying for the Part 107 exam, I came across information about flying a drone around an airport. I thought, "I do photography. I have no interest in flying around an airport. Why do I need to know this information?" Sometimes you'll feel baffled when you realize all the FAA wants you to know, and you may have already felt that way from just reading this book!

But here's an answer to that why-do-I-need-to-know-this question: "What if a Realtor hired you to shoot a building for sale that was very near an airport?" This answer brings possible scenarios into focus. After all, some airport regions are rather large. If you're working in one of those areas where commercial planes are taking off and descending, you obviously do need to know what's going on in the air around you!

TIP

This chapter covers the FAA rules for flying your drone in the vicinity of airports and familiarizes you with airport traffic patterns. Don't worry: I'm not prepping you to become a commercial pilot, but this chapter will help you fly safely in an area where the skies are crowded.

Reading Chart Supplements

The past several chapters focus on reading sectional charts, and for good reason. You'll see sectional charts on the Part 107 exam, but you also need to know how to read them for practical purposes.

The same is true for airports. You may have no plans to fly around an airport at the moment, but when you do, you'll need to be able to get some info about that particular airport. That's what chart supplements are for. *Chart supplements* are simply text-based information about a particular airport that is updated frequently, so you can rest assured that you have the right information.

You can access chart supplements at http://skyvector.com. Just find the airport you want to get information about and click the icon on the airport. For example, to look at a small airport in Mena, Arkansas, called the Mena Intermountain Municipal Airport (MEZ), you can locate the airport on the map and click the icon on the airport (shown in Figure 13-1), or you can just search http://skyvector.com for the airport.

Airport icon

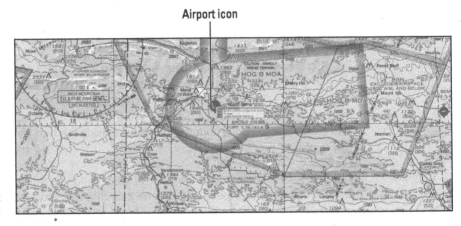

FIGURE 13-1:
Locate the
desired airport
on the map.

After you click the airport icon, information about the airport appears (called a chart supplement), organized by categories, as shown in Figure 13-2. A lot of this information is for crewed aircraft pilots and doesn't particularly apply to drone pilots. However, some of the information can be helpful to you:

>> **Communications:** You'll find a section on the chart supplement for communication frequencies for a particular airport. For example, say you need to fly near an airport and you want to monitor UNICOM for that airport. The chart supplement will give you the frequency (122.8 for MEZ, for example).

>> **Traffic pattern:** Most all airports use a left traffic pattern (more on traffic patterns later in the chapter), but the chart supplement will confirm it. If you're flying a drone near an airport, you need to know how crewed aircraft traffic approaches the airport and how it takes off so that you can have a clear sense of the traffic direction.

>> **Control tower:** The chart supplement tells you whether the airport has a control tower. (Most airports don't.) You can check out more information about towered and nontowered airports in the next section of this chapter.

TIP

>> **Notes and additional information:** A helpful aspect of chart supplements for drone pilots is notes or an "additional information" section, which some chart supplements have. For example, the MEZ chart supplement notes rapidly changing elevation due to nearby mountains as well as migratory birds and the presence of deer. These kinds of notes may help you be more aware of possible dangers to drone piloting in the area as well.

No control tower

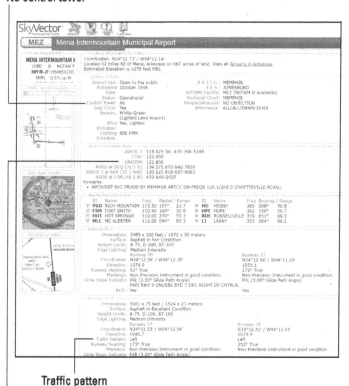

FIGURE 13-2:
Chart supplement airport information.

Traffic pattern

Airport communications

Understanding Towered and Nontowered Airports

Most think of an airport as having a control tower with air traffic controllers (ATC) who communicate with an airplane for takeoff and landing. It's true that bigger airports have a control tower, and airplanes maintain two-way radio frequency with the tower (and the pilots have to do whatever ATC tells them).

What you may not realize is that only about 500 airports in the United States have a control tower. Shockingly, the other 20,000 airports do not. These other airports are medium-sized to rather small regional airports. The good news is that planes take off and land safely every day at both towered and nontowered airports.

As a drone pilot, it's important for you to understand that most airport traffic in the United States does not have two-way communication with a control tower. However, at nontowered airports, pilots typically self-announce their intentions regarding landing and taking off. Because there are so many small airports around the country, the odds are good that you'll be piloting near one at some point for one reason or another. In these cases, you want to know if a plane is landing or taking off. This is one way in which the chart supplements described in the previous section are helpful.

For example, GLE's (Gainesville Municipal Airport) chart supplement tells you that you can monitor UNICOM and CTAF frequencies at 123.0, as shown in Figure 13-3. That's important because that's how you can hear pilots self-announce their intentions if you're flying in the vicinity.

TIP

Many drone pilots end up purchasing a VHF radio so that they can connect to these local frequencies and know what's going on in the airspace around them. You can find out more about radio communications in Chapter 10.

UNICOM and CTAF frequency

FIGURE 13-3:
An airport's communication frequencies are provided on the chart supplement.

Exploring Airport Traffic

The FAA wants drone pilots to have a very good understanding of airport traffic, runways, and how traffic approaches a runway and takes off. The goal, of course, is safety. If you understand how airport traffic works, you can make good piloting decisions that prevent accidents. You would never want your drone to run into a crewed aircraft, after all. So how does airport traffic work? As you might guess, it's complicated, but I tell you only what you need to know.

The rundown on how runways work

Runways are organized according to compass headings, so to understand a runway, you first have to review the compass, shown in Figure 13-4. A compass shows you where north, south, east, and west are, and the compass headings are represented by degrees, as follows:

>> **North:** 360 degrees

>> **East:** 90 degrees

>> **South:** 180 degrees

>> **West:** 270 degrees

FIGURE 13-4:
Compass
directions
and degrees.

Airport runways are numbered between 1 and 36, and those runway numbers correspond to the numbers on a compass. You just drop the zero on the compass heading for the runway number, meaning that 90 degrees is runway 9. The runway number tells you the direction of the runway, and therefore the direction of the planes landing and taking off on that runway. So, for example, if you see runway 14 at an airport, that number overlays a compass's magnetic direction of 140 degrees and tells you the direction that planes are landing and taking off, which is southeast in the example shown in Figure 13-5.

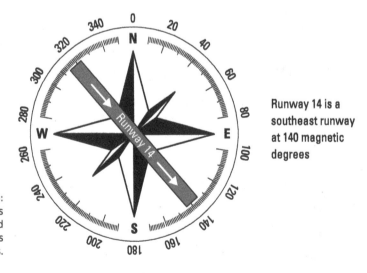

Runway 14 is a
southeast runway
at 140 magnetic
degrees

FIGURE 13-5:
Runway numbers
correspond
to compass
headings.

TIP

It may seem odd that all traffic flows the same way on a particular runway, but think of that runway as a one-way street. In order to avoid collisions, all traffic flows the same way. A plane lands and taxis off the runway, and then perhaps another plane taxis onto the runway and takes off, but it takes off in the same direction. If the traffic didn't flow in this manner, there would be a great chance of midair collisions with planes taking off and landing.

TIP

It's not unusual for both directions of a runway to be used, but, of course, not at the same time. So in this example, the other end of the runway will have the appropriate compass heading as well. For example, in Figure 13-5, the opposite of runway 14 is 32. Each will be considered a different runway (14 and 32), but they're the same physical runway.

TIP

It's also important to note that each airport is configured a bit differently: Some airports use only certain runways for takeoffs but may use a parallel runway for landings. Some use a runway in only one direction because of obstacles (such as a runway near a mountain). Often, runways are built in certain compass heading to avoid wind. Just keep in mind that each airport is configured differently.

To clarify with one more example, say that you hear airport traffic chatter talking about runway 22. What direction is this runway? In other words, as planes touch down on the runway, what direction are they going?

The answer: southwest.

If you get confused, just make a quick sketch of a compass, and then overlay the runway on the compass. Doing so will help you get the right directional heading, as shown in Figure 13-6.

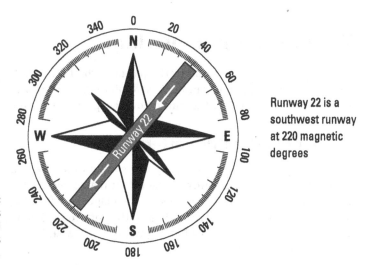

Runway 22 is a southwest runway at 220 magnetic degrees

FIGURE 13-6: Runway numbers correspond to compass headings.

TIP

If you're taking the Part 107 exam, you'll get scratch paper at the testing center. Make sketches with questions related to airports and directions! Your sketches will help you keep from getting confused.

Understanding runway markings

For the Part 107 exam, you also need to understand the basics of runway markings and what they mean. That requirement seems sort of odd, considering that you won't be landing your drone on a runway. Nevertheless, they're not complicated. The following list explains the runway markings, but be sure to check out Figure 13-7, which will help these explanations make sense:

- » **Unusable area:** Yellow chevron marked areas are unusable parts of the runway, except in cases of emergency.

- » **Displaced threshold:** This is the area leading up the landing portion of the runway. It may be used for taxiing and even takeoff, but not for landing.

- » **Threshold:** This is the beginning of the landing, or touchdown area of the runway.

- » **Runway number:** The runway number is painted on the runway and is easy for pilots to see.

- » **Centerline:** This marking, similar to a highway marking, just helps pilots see the center of the runway.

- » **Aiming point:** These big, painted block areas are the ideal location for planes to touch down, although in reality, touchdown takes place anywhere in this general area.

Knowing how planes land

Most airports follow the same traffic pattern. The FAA wants you to know this traffic pattern so that you can make safe piloting decisions with your drone. In other words, knowing how the traffic flows keeps you out of the traffic!

The left-hand traffic pattern

Most all airports follow a *left-hand*, also simply called *left*, traffic pattern. This means that all planes only turn to the left. It makes sense when you think about it. The pilot's seat is on the left side of the plane, so left turns give the pilot the best visibility. Also, left turns create a traffic pattern so that all planes are doing the exact same thing. That's what you want — it's what keeps planes from running into each other!

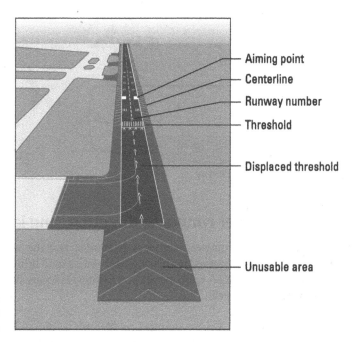

Aiming point
Centerline
Runway number
Threshold
Displaced threshold

Unusable area

TIP

Sometimes traffic directions are changed for very bad weather or other serious problems, but just remember, globally, that planes landing at and taking off from airports turn left.

The "legs" of landing

Planes that are landing at an airport follow what are called legs. *Legs* are simply different parts of the landing pattern, as follows (and shown in Figure 13-8):

» **Upwind leg:** This flight path is parallel to the landing runway in the direction of landing. This leg is used after takeoff to gain altitude in order to move out of the landing/takeoff pattern or to enter the crosswind leg.

» **Crosswind leg:** This flight path is at a right angle to the landing runway from the takeoff end. The leg provides separation from the runway and gives pilots the ability to move onto the downwind leg.

» **Downwind leg:** This flight path runs parallel to the landing runway in the opposite direction of landing. Basically, you can think of the downwind leg as the plane entering the landing traffic. Pilots often enter the downwind leg on approach for landing, at a 45-degree angle.

» **Base and final leg:** This flight path is at a right angle to the landing runway, extending from the downwind leg to the intersection of the runway centerline. This leg is considered the final approach for positioning and landing.

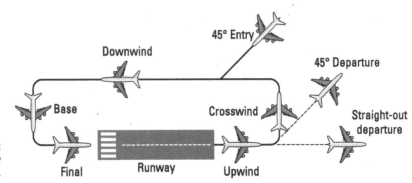

FIGURE 13-8:
Runway
leg pattern.

Bringing together runways, direction, and landing

After you understand runways, compass headings, and the workings of traffic patterns at airports, as covered in this chapter, you need to bring all this information together. What the FAA really wants you to know is where the traffic is around the airport landing pattern.

Consider this scenario: You're monitoring airport traffic and you hear a pilot declare a position of "midleft, downwind to runway 14." Where is the aircraft right now? North, south, east, or west?

The way to solve this question is to sketch it out, at least until you can mentally draw it. Starting with the runway, you can put the runway on a compass heading to literally get your bearings. As you can see in Figure 13-9, runway 14 corresponds to 140 degrees on the compass.

FIGURE 13-9:
Compass heading
for runway 14.

Next, you can add the leg pattern to this drawing to figure out where the airplane is. Remember, all turns are left, so applying the leg pattern shows that "midleft, downwind" means that the plane is flying parallel, but opposite, to the runway, which will bring the pilot to the final approach. Currently, the plane is in the northeast (see Figure 13-10).

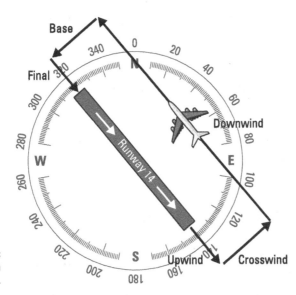

FIGURE 13-10:
The plane is in
the northeast.

Try another example to make sure that you've got it. Say that you hear a pilot announce "midleft, downwind to runway 2." Where is the airplane?

The plane is in the west, as Figure 13-11 shows. Always remember that every turn has to be to the left.

REMEMBER

The weather can have an impact on the traffic pattern, especially when the air is denser (think hot, humid days). This dense air makes takeoff harder, and climbing to desired altitudes takes longer. The landing also takes more space. A traffic pattern is therefore adapted as needed to accommodate weather conditions, making the pattern longer, denser, shorter, and so on.

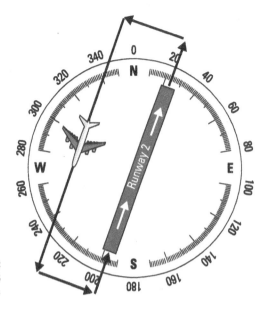

FIGURE 13-11:
The plane is
in the west.

Understanding Right of Way

The concept of right of way means that in a traffic situation, one vehicle has the "right" to go a certain way and a second vehicle yields that right. This basic concept is used for automobile traffic on roads to avoid collisions, and the same is true for vehicles in the skies.

For drone pilots, this concept is pretty straightforward. You have to give right of way to everything else around you. In fact, it's unlawful for you to fly your drone in any way that interferes with operations at an airport, helipad, or seaplane base. A part of this rule is to understand that you have to yield the right of way to every other aircraft. Global rules state that an aircraft in distress always has the right of way, but this doesn't include drones in distress!

REMEMBER

Along with right of way is the concept of see and avoid. This concept comes back to the idea of maintaining situational awareness that I mention throughout (and discuss in Chapter 16), and knowing exactly what's happening around your drone at all times.

If the FAA seems over the top in requiring all this information about airplanes, runways, and airport communications, just remember that the idea is to help drone pilots perform jobs near airports but have good situational awareness so that they can avoid accidents. In the end, safe flying is always the goal.

4

Flight Operations and Safety

IN THIS CHAPTER

» **Exploring air masses**

» **Checking out convection currents**

» **Flying in the wind**

» **Flying in humidity and fog**

» **Reading METAR and TAF reports**

Chapter **14**

Piloting in Various Weather Conditions

O f all the potential flying problems a pilot faces, the weather is the greatest. From drone piloting to commercial airline piloting, the weather can have a significant impact on flight safety. Have you ever sat in an airport with a flight delay due to the weather? Frustrating, but there's a good reason for it! If you're in the air, the weather can be a serious danger to property and life.

Although a drone has no human occupants, the weather can cause flight dangers and related problems. A responsible drone pilot must be well aware of potential weather problems and how those problems can impact a flight.

From the safety of your drone to the safety of people and property near you, it's extremely important for you to know a thing or two about the weather.

TIP

The Part 107 exam expects you to know some things about the weather as well. Read on!

Exploring Air Masses

The atmosphere is basically made of different kinds of *air masses*, which are just big areas of air. Air masses can vary from one another in their characteristics and the weather phenomena they produce. Land, water, and temperature all affect air masses, which is why the air mass you're likely to experience in Nevada is different from the air mass you experience in Hawaii. This variation matters because your drone's performance is affected by different kinds of air masses, leading to different kinds of weather.

Different air masses appear over large surface areas around the globe known as *source regions*. These source regions can span thousands of miles across the surface of the earth, and the air mass above them can reach up to ten miles high. These giant air masses absorb the features of the source region, and that's why the earth has various types of air masses, such as

- **Maritime:** Humid air masses that form over large bodies of water
- **Arctic:** Very cold air masses that appear in arctic regions
- **Tropical:** Warm, often humid air masses that appear in tropical regions
- **Continental:** Dry air masses that appear over large land masses

TIP

Talk about different kinds of weather and air masses commonly involves the word *convection*, which is the vertical movement of heat and moisture from the ground up. As the ground heats up during the day, that heat and moisture rises. This rising motion creates different kinds of weather phenomena, such as thunderstorms.

Connecting cloud types with weather systems

Air masses can contain varying kinds of clouds, and you need to know about the impact of cloud types on weather. You may be familiar with some of these, but here's a quick review (and see Figure 14-1):

- **Cumulus:** These clouds are what you think of when someone says the word *cloud*. They're the big, white, fluffy shapes that appear on nice sunny days as the sun heats the ground (resulting in *diurnal convection*). You can think of them as "fair weather" clouds.
- **Stratus:** These clouds hang low in the sky, and you often see them on overcast days. You may also see light drizzle when these clouds are in the sky.

- **Stratocumulus:** These clouds are low and dark, and you may see blue sky between them. They're basically groupings of stratus clouds that spread across the sky, forming when there's weak convection.

- **Altocumulus:** These mid-level clouds appear as white or gray patches; they may also show up in groupings or as round or banded clouds.

- **Cirrus:** These clouds have a feathery appearance and occur high in the sky. They usually appear during good weather and stable conditions.

- **Cirrostratus:** These thin clouds appear high in the sky and often precede weather systems with rain or snow.

- **Cirrocumulus:** These clouds often appear as streaks high in the sky, typically before rain.

- **Altostratus:** These mid-level clouds are often gray or blue-gray and may obscure the sun. They often appear before rain.

- **Nimbostratus:** These dark, low clouds bring steady rain or snow. They usually appear during longer periods of precipitation.

- **Cumulonimbus:** These large storm clouds bring thunderstorms, heavy rain, lightning, and possibly hail. They can develop quickly and almost always bring severe weather.

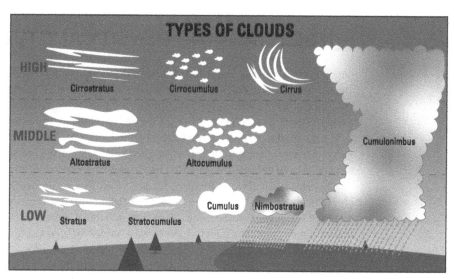

FIGURE 14-1: Types of clouds.

TIP

As a drone pilot, which two clouds are the most troubling? You guessed it! Nimbostratus and, especially, cumulonimbus. These clouds bring rainstorms and other severe weather.

Getting current on stable and unstable air

In aviation, you often hear the terms *stable air* and *unstable air*. These terms can be a bit confusing, but you need to know what they mean because both impact you as a pilot.

Convection currents move air in a vertical direction. When the air is stable, little movement occurs. With stable air, you see stratiform clouds, fog, and, often, steady precipitation. The air is stable (not moving), but usually you have poor visibility with haze, fog, and smoke. Because the air isn't moving in a vertical direction, the lower-level air collects dust particles. The combination of all these elements means that you can't see well in stable air, which impacts the safety of drone piloting missions.

Unstable air has convection currents at work moving the air in a vertical manner. These currents create air that's unstable, leading to turbulence. With unstable air, you see cumuliform clouds and showery precipitation. The visibility is generally good, but the unstable air flow and turbulence can certainly affect any drone or aircraft.

REMEMBER

When you hear the term *stable air*, think poor visibility. When you hear the term *unstable air*, think good visibility, but turbulence.

UNDERSTANDING DENSITY ALTITUDE

Another term you hear in aviation is *standard day*, which means that at sea level, the temperature is 59 degrees Fahrenheit and the air pressure is 29.92 inHg (inches of mercury, or 1013 millibars). So think of these numbers as the standard. Another term to know is *density altitude*, which describes the density of the air as you're flying relative to the standard day. As described in Chapter 1, the drone's propellers use the air to create lift, and the less density of air there is at a particular altitude, the more work the drone has to do to fly because the air is thin. This is where the concept of above ground level (AGL), explained in Chapter 6, comes into play. If you fly your drone 200 feet in the air from the beach, the density altitude will differ significantly from flying your drone at 200 feet in Denver, Colorado. The AGL in Denver is much higher than sea level, so the air is thinner. The drone therefore has to work harder to fly, reducing the life of the battery. As density altitude increases, the performance of your drone decreases. Higher density means thinner air, whereas lower density means thicker air. Just remember that if you fly in regions high above sea level, you can expect decreased drone performance.

Piloting through Convection Currents

You should use a weather forecast before you fly, which can certainly help you make piloting decisions. However, one thing to keep in mind that the weather forecast isn't going to mention is convection currents and how those currents can impact a flight.

Convection currents are just horizontal air currents. As the sun heats the earth's surface, that heat rises, creating a current. However, convection currents can vary a lot based on what the sun is hitting on the surface. Water, plants, grass, and trees absorb the heat and can create downward currents, whereas sand, rocks, pavement and other reflective surfaces push the heat back up into the air. You've probably seen heat rising off a parking lot in the middle summer. This is effectively a convection current pushing the heat back up.

When you fly the drone, these downward and upward currents affect the stability of a horizontal flight, putting more upward or downward pressure on the drone, depending on what you're flying over, as you can see in Figure 14-2.

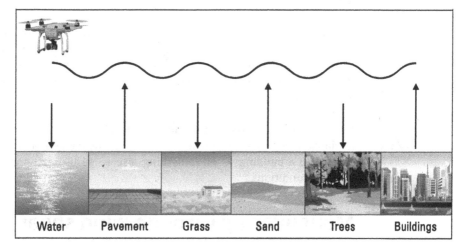

FIGURE 14-2:
Convection currents can impact drone flights.

REMEMBER

Convection currents can create turbulent flights at lower altitudes, depending on what you're flying over. Just remember that on warm days, surfaces that reflect the sun's heat will create currents that push your drone upward, whereas surfaces that absorb heat create currents that can pull your drone downward. These convection currents at lower altitudes can certainly degrade your drone's performance and make maneuvering it more difficult.

Dealing with the Wind

Of all the weather-related problems you may experience as a drone pilot, the wind is often the most difficult. Wind problems can be hard to predict, greatly degrade performance, and even cause you to lose control. I've personally had only one near accident with a drone, and it was due to too much wind degrading the battery life faster than I thought it would. I barely got the drone back to me and landed before running out of battery. So don't underestimate the wind!

Aside from the problem of battery degradation on windy days, you need to be aware of a couple of other issues concerning flying in the wind. Read on.

Flying with obstructions in mind

When you pilot a drone, you may not be that far off of the ground. Perhaps you're inspecting some kind of structure or photographing a construction site. However, even if you're flying relatively low, it's important to note that larger, nearby structures on the ground affect the wind and can create unstable air flow and wind gusts.

For example, say you're flying in an urban area with several tall buildings around. When there's some wind, those buildings affect wind flow because the wind is getting blocked and flowing around those buildings. What happens when the air flows around those buildings and over the tops of them can be unpredictable, creating gusts and sudden changes in direction.

WARNING

Just remember that flying a drone around tall buildings in windy conditions can be dangerous. It's easy to lose control of the drone and be unable to prevent a crash into a building — or something else — so use extreme caution in these kinds of conditions.

Maneuvering through wind shear

Wind shear is a sudden and violent change in wind speed and direction. Wind shears can occur horizontally or vertically and have been the cause of several aviation accidents.

Wind shears occur over short distances, but of course, if your drone (or even an airplane) is flying through that short distance, the wind shear can greatly affect maneuverability, takeoff, and landing. Wind shears typically occur during thunderstorms or with passing frontal systems, so it's always a good idea to consider these weather issues before you fly.

Staying up in downdrafts

If wind is traveling over a mountain, you can encounter a downdraft, which can cause your drone to crash in severe cases. You don't want that, so here's the skinny: As wind flows over a big ground structure, the air movement is stable as it flows up the side of the structure. It's the backside of the structure (leeward) that's problematic. The air is pushed back down toward the ground from the higher altitude it gained while moving over the structure, as shown in Figure 14-3. This downdraft creates unstable currents pushing downward, and if the air is rapidly moving downward, it's going to push your drone in the same direction as well. As with human-made structures, use extreme caution flying in mountainous areas on windy days.

FIGURE 14-3: Downdrafts create unstable, downward air flow.

Watching Out for Humidity and Fog

Although wind is a primary weather consideration when you pilot your drone, humidity and fog also come into play as safety considerations. You don't need to become an expert on either topic, but you should know a few things about humidity and fog, and how they can impact safety.

Working harder in low and high humidity

The atmosphere contains water in all three forms: liquid, solid, and gas. For this reason, we experience rain, snow, hail, fog, and everything in between. Water in the form of gas is called water vapor. *Humidity* refers to the amount of water vapor in the air, and the temperature of the air directly affects how much water vapor the air can hold. Every 20-degree increase in air temperature doubles the amount of water vapor the air can hold, and the reverse is true as well. That's why the air on cold winter days is dry and the air on hot summer days is damp, or humid. Water vapor is added to the air as water evaporates from the ground.

With that said, there are a few concepts you should know for drone piloting. First, *relative humidity* refers to the amount of water vapor the air can hold at any given time. This value is expressed in a percentage. For example, a relative humidity of 60 percent means that the air is currently holding 60 percent of its water-vapor capacity. The *dew point* is the temperature at which the air becomes saturated with water vapor (100 percent humidity). At the dew point, condensation begins to form, and you see dew and usually fog.

REMEMBER

Understanding humidity matters for a drone pilot because in thin air with very low humidity, the drone has to work harder to fly, but the reverse is true as well. In very humid conditions, the air is thick, and the drone also has to work harder to push against it. So if you're piloting in very humid conditions, you should expect that your drone's battery will get depleted more quickly.

Figuring out fog and visibility

Fog is a very low-lying cloud that occurs when the air becomes saturated with water vapor. The air can't hold any more water, but that water has to go somewhere, and in most cases, it turns into fog. Fog affects a pilot's visibility, so the FAA considers fog a safety issue that pilots should be prepared to contend with. That said, if you take the Part 107 exam, you may see a scenario question asking you about the kind fog you might expect in a certain situation. Here are the kinds of fog:

>> **Radiation:** Radiation fog forms in low-lying areas (such as a valley) when the dew point is reached and the ground cools after sunset. This fog will hang low to the ground and then dissipate as the sun warms things up the next day.

>> **Advection:** This fog often occurs when southern, moist air meets cool moisture on the ground.

>> **Precipitation:** This fog forms when rain falls through cold air. This fog commonly occurs with warm fronts.

>> **Upslope:** This fog tends to form on mountaintops as moist wind pushes air toward the mountain, causing cooling as the air flows toward the top.

>> **Steam:** This fog forms over lakes when cold, dry air moves over the lake water, which is warmer.

>> **Valley:** This dense fog forms in valleys due to moisture from the soil cooling to near the current dew point.

>> **Freezing:** When temperatures fall below freezing, this fog can form and cause freezing droplets of water to form on surfaces. Think of this fog as "foggy drizzle."

>> **Ice:** This fog forms only in polar and Arctic regions in extremely cold temperatures.

REMEMBER

Foggy conditions reduce visibility, so they always pose a danger, regardless of the kind of fog you experience. It's good to understand the kinds of fog and when it's likely to occur so that you can project possible flight problems and delays.

Understanding METAR and TAF Reports

The Aviation Weather Center provides two different kinds of weather reports. You can access the current reports on a region around an airport anytime at https://aviationweather.gov/data/metar/. You can access two different kinds of reports: METAR and TAF. You can check out the details of these reports in the following sections, and you need to understand METAR and TAF reports for the Part 107 exam.

METAR

An *Aviation Routine Weather Report (METAR)* is an hourly report noting ground weather in a particular area. METARs can be issued more often than hourly if a significant weather event is occurring. These more frequent reports are called special METAR (SPECI) reports. Here's what a METAR report looks like:

```
KDFW 021553Z 33022G29KT 10SM OVC023 13/08 A2987 RMK A02 PK WND
32029/1458
```

If that looks like a nightmare, just know that it's not as bad as it first appears after you know how to read it. The best way to do that is to break it down into its parts because that's what a METAR is — a report with several pieces of information crammed together. Here are the parts:

>> **KDFW:** This is the *call sign* of the reporting station. Typically, it's an airport code with a K in front of it. KDFW is Dallas Ft. Worth International Airport.

>> **021553Z:** This is the day and time of the report (in Zulu [also known as UTC] time). This report was generated on the second day of the month (02) at 1553 Zulu time. Zulu time is synonymous with UTC (Coordinated Universal Time), which is a prime time zone used by the military and aviation. It remains the same regardless of Daylight Saving Time.

>> **33022G29KT:** This part of the report refers to the wind speed. The first three numbers is the wind direction (330) and the last two are the speed (22). *G* means "gusts" up to 29 knots (KT). As you can see, this is just a bunch of information put together in one group of numbers and letters.

>> **10SM:** This represents visibility. At DFW, the visibility at the time of this report was 10 statute miles (a unit of measurement used to determine distances on land). If visibility is poor, this number can be expressed in fractions, such as 3/4SM (three quarters of a mile visibility).

>> **OV023:** These characters refer to the weather and sky conditions. At DFW for this report, it was overcast at 2,300 feet. There are several different weather abbreviations, which I point out in the Tip following this list.

>> **13/08:** Temperature and dew point. The temperature is 13 degrees Celsius, and the dew point is 8 degrees Celsius.

>> **A28987:** This is the altimeter setting.

>> **RMK:** This set of letters refers to a remark. A METAR report may contain remarks about additional weather issues. This report contains a remark about peak wind speeds and times.

TIP

Obviously, if you access a METAR, you probably want to understand the weather conditions, so those codes are most helpful:

>> **BKN:** Broken

>> **OVC:** Overcast

>> **BR:** Mist

>> **SH:** Showers

>> **RA:** Rain

TIP

The confusing aspect of METAR codes is that they may be pressed together. For example, RAB15 means "rain began 15 minutes past the hour." Want to more about all of the codes and what they mean? Check them out at https://www. weather.gov/media/wrh/mesowest/metar_decode_key.pdf.

TAF

A *Terminal Aerodrome Forecast (TAF)* is a report of a five-mile radius around a specific airport. Thank goodness, a TAF report uses the same basic codes as a METAR, but it just provides very specific information about what's happening, weather-wise, around the airport. You access a TAF report in the same place as a METAR, at https://aviationweather.gov/data/metar/.

TAF reports are not generated hourly, but typically about four times a day, so they're not as current as METAR reports (which are updated hourly or more if needed).

Here's a sample TAF report for DFW airport:

```
KDFW 021726Z 0218/0324 32015G25KT P6SM SCT030 FM030100 33015KT
P6SM SCT050 FM030500 290009KT P6SM SKC FM031500 33013KT P6SM SKC
```

Here's what each part of this report means:

>> **KDFW:** This is the station identifier, which is DFW airport.

>> **021726Z:** This report was prepared on the 02 day of the month at 1726 Zulu time.

>> **0218/0324:** The report is valid from the second day of the month at 1800 Zulu until the third day of the month at 2400 Zulu.

>> **32015G25KT:** This section contains wind speed, visibility, and other issues related to weather. This location currently has wind gusts (*G*) up to 25 knots (KT).

>> **P6SM:** Visibility to five square miles.

>> **SCT030:** Possible scattered thunderstorms until 0300 Zulu.

The rest of the report gives you ranges of visibility from one time to another (FM means "from") for wind speed and overall weather and visibility (SKC means skies are clear). Of course, what you see here will vary greatly depending on the weather conditions.

When you need to carefully look at weather conditions near a specific airport, a TAF can help, but don't forget that you can also access information hourly or more often using a METAR. For this reason, METARs are often more practical for drone pilots who are operating in the general area of a specific station identifier.

REMEMBER

You probably won't use a METAR enough to make reading one quick and easy. But if you need this resource, don't forget to grab the code key at https://www.weather.gov/media/wrh/mesowest/metar_decode_key.pdf. This key will help you quickly decipher the report. Also, if Zulu time mixes you up, check out this quick Zulu-to-standard-time converter at https://www.timetemperature.com/zulu-converter/zulu-converter.php.

Tips for Flying in Bad Weather

The best tip is not to fly in bad weather. However, avoiding it is not always possible, especially because bad weather can sometimes sneak up on you during a drone piloting mission. If that happens, here are some important, global tips to keep in mind.

» **Humidity and wind affect drone performance.** Always keep that in mind. The batteries will not last as long as you might think, so be very aware of this limitation when you're flying in windy or very humid conditions. Also, thin air, such as in mountain ranges, affects performance as well.

» **Think carefully about thunderstorms.** It may seem obvious that you can't fly in a thunderstorm, but what about the area around the thunderstorm? Here's the thing: Rain and possible hail can fall outside the area where the thunderstorm is happening. General aviation advice is to stay 20 miles away from a current or developing thunderstorm.

» **Be careful in below-freezing temperatures.** If the temperature is below freezing and water vapor is in the air, ice can form on your drone as it flies through this water vapor. Ice equals weight, which equals problems, so use a lot of caution if you must fly in these conditions.

WARNING

» **Hail is sneaky.** Hail can fall miles away from a thunderstorm, a cold front, or in various other conditions. This frozen water vapor can sneak up on you, especially in the early spring when the weather can be volatile. Hail can destroy a drone in flight, so be careful and remember that hail is sneaky!

» **Use METAR and TAF reports.** If you're flying near an airport (make sure you have FAA permission, if necessary), a TAF can be helpful. Otherwise, use a METAR for the best information. Although frustrating to read, METARs give you detailed information that helps you plan a piloting mission. The weather app on your phone isn't going to give you the details you need, so grab a METAR when you need it!

Chapter **15**

Flight Emergency Procedures

No matter how often you fly or how much you practice, eventually something is going to happen that you will consider a flight emergency. For drone piloting, a flight emergency occurs when something can potentially cause the drone to crash, or you to lose connectivity with the drone. Either way, you risk losing the drone, or in the worst-case scenario, damaging property or injuring a person.

The FAA wants you to know how to respond to flight emergencies, and for all practical purposes, so do you! The good news is that modern drones are very stable and fly very well, and there isn't a ton of information you need to know. However, when you do encounter an issue, a few key pieces of information can help you avoid an accident. In this chapter, I tell you about common issues that lead to problems and how to avoid them, as well as how to deal with an emergency.

This chapter also discusses some built-in safety features that your drone probably has. Be sure to check out Chapter 2 for additional details so that you can find out what features are included with your drone.

Looking at Common Drone Emergency Causes

When you think of a flight emergency, you probably think of a human-controlled aircraft. After all, an emergency on a human-controlled aircraft can be a life-or-death situation, and that is, of course, true. With a drone, however, it's easy to think, "It's a drone. How big of an emergency can it be?"

That's a thought everyone needs to expel, and for good reason:

>> **Damage:** It goes without saying that a drone crash can cause property damage or injury to a person. However, if you have a drone weighing under 250 grams, you may be tempted to think that the probability of such an outcome is minimal. That may be true, but that thinking can continue even when you upgrade to a bigger drone. Some of the prosumer (an overlap between consumer and professional level) drones are, in fact, large and heavy. Also, an object falling from the sky doesn't have to weigh a lot to cause significant damage to a car or person. Don't underestimate it!

>> **Money:** If your drone damages someone's property, that's going to cost you. You'll also need to replace your drone, which isn't cheap! Keeping in mind the financial implications of a drone accident may help you take the prospect seriously.

The first thing you should know is what most often causes drone emergencies. Armed with this information, you can watch out for these pitfalls! Here are the top issues.

Lost link

A lost-link emergency occurs when the drone loses radio connectivity to the controller. There are some specific procedures for this problem, which I cover in "Understanding Lost-Link Procedures," later in the chapter.

Weather

Sudden weather changes can create some of the greatest problems for drone pilots. Depending on where you're flying, the weather can also change rapidly and unexpectedly. One time I was flying in Utah with some wind, but not enough wind to cause any problems. I finished my current photography shoot, landed, and packed up. Ten minutes later as I was driving away from the location, a massive windstorm came out of nowhere. It was hard to even drive a car with so much sand blowing everywhere. I remember thinking, "You barely escaped that!" Nothing in the weather report could have alerted me.

REMEMBER

Do everything you can to check the weather forecast, but also know that weather can be volatile in different locations and at different times of the year. Bad weather can quickly cause you to lose control of your drone and crash.

Batteries

Always make sure the drone and controller batteries are fully charged before you fly, but also know that altitude and wind can have a great impact on battery life. I would say wind is one of the worst. Wind can put a big strain on your drone, causing it to use much more battery power than you think just to remain stable. No matter what conditions you find yourself in, never forget to keep a close eye on the battery life. And, if it's windy, it will take more battery power to get the drone back to you and land safely than you might think!

Birds

The second time I flew a drone, a bird tried to attack it. I managed to get away, but honestly, it wasn't because of my skill. It was just luck! It's important to note that birds can be a big problem. Some birds will avoid drones and others become aggressive, especially if you fly too close to a tree with a nest. Some birds are very territorial, and predatory birds will attack drones "just because." Also, if you fly around the ocean, note that seagulls can be particularly problematic.

Depending on where you live and fly, you may have more or fewer problems with birds, but don't underestimate their ability to crash your drone.

To avoid birds, first do your homework and know what kinds of birds you may encounter in the area you're flying. Also, it's helpful to know about migration patterns and whether you'll encounter a lot of birds during a certain time of year.

TIP

With that in mind, if a bird tries to attack your drone, you want to fly away and up at the same time. Most birds attack from behind and slightly above their prey, as shown in Figure 15-1. You want to both get away but also above the bird and out of the attack pattern (as long as you don't fly above the 400-foot maximum height). Quick thinking and quick joystick movements are needed here!

FIGURE 15-1:
Birds will
attack drones.

Africanway/Getty Images

Overdependance on obstacle avoidance

Your drone may be equipped with some level of obstacle avoidance. This system works using sensors, and your drone may have more or fewer sensors depending on the model (and how much money you paid for it).

These sensors and systems are great, but they're not fail proof. Many drone pilots become overly dependent on them and stop flying defensively, thinking the collision system will always protect the drone. However, nothing is as good as your eyes. Drone sensors may not see everything, especially small tree branches, power lines, and other small obstacles. So, use the obstacle avoidance system, but don't depend on it. Use a combination of obstacle avoidance and your eyesight.

TIP

Remember — most drone obstacle avoidance systems do not work in sport mode. I encourage you to use sport mode only when you're sure you can fly safely without obstacle avoidance.

Carelessness

Unfortunately, drone pilots may be the greatest cause of emergency situations. A drone is just a machine, and the human factor contributes significantly to that machine's safety and ability to remain in the sky.

Often, pilots become careless or do not take seriously the implications of a crash. For this reason, the FAA identifies pilot attitudes and thinking patterns that frequently lead to crashes. See Chapter 16 for details.

Practicing Foundational Emergency Responses

The previous section of the chapter goes over the most common causes of emergency situations. So when potential emergency situations are brewing, what do you do? Here are some foundational emergency responses:

>> **Yield right of way.** The FAA expects drone pilots to yield right of way to any other aircraft. That includes planes, helicopters, hot-air balloons, and even other drones. Too often, a drone pilot may think, "I was here first." That's wrong thinking. The safe way to think and avoid accidents at the same time is to get away from anything else in the sky.

TIP

>> **Avoid distractions.** Drone piloting can be fun with other people. The problem is, other people create distractions. Conversations, interruptions, or just nearby people walking around can distract you from your mission and potentially cause an accident. Some people, both friends and strangers, are naturally curious about drone piloting and will approach you to discuss it — while you're in-flight! For this reason, the FAA wants drone pilots to wear a safety vest, such as the one shown in Figure 15-2. This vest can alert people to stay away. In fact, you can even purchase safety vests with text printed on the back saying "FAA Drone Pilot. Do not disturb" or something similar. These products can be helpful, depending on where you're flying.

FIGURE 15-2: Safety vests can help reduce human distractions.

Magove/Getty Images

>> **When in doubt, land.** One of the best emergency responses you can practice is the attitude of "land the drone." Here's what I mean: Some drone pilots try to fight it out. Maybe the weather is not looking good; maybe there are too many obstacles; maybe you feel distracted, anxious, or stressed. Landing your drone in the middle of a mission isn't a failure. Landing gives you a chance to reassess the situation and make decisions without being rushed. So, when in doubt, just land the drone and think things through. You can always take off again if conditions appear to be safe.

Understanding Lost-Link Procedures

At one time or another, virtually every drone pilot experiences a "lost link." The FAA defines *lost link* as a loss of command and control between the drone and ground station. That basically means that your remote control (RC) loses connectivity with the drone while the drone is in flight. Lost links can occur in one of two ways (or both at the same time). First, you can lose the uplink, preventing the RC from sending control information to the drone. Or you can lose the downlink, which keeps the drone from transmitting information to the RC (such as telemetry, video, sensor data, and everything else).

Knowing why lost links occur

Lost links can occur for a number of reasons. Here are the most common:

>> **Transmission range:** The RC and drone have a maximum transmission range, which varies based on drone make and model. However, the maximum transmission range doesn't mean you'll get that full range. Various things, such as buildings, topography, and even weather, can interfere with transmission range. In fact, you can fly the same drone mission two days in a row and not achieve that same range each day. So you can lose the link between the RC and drone if the drone gets out of range, and you can't be entirely sure what that range is. Pay close attention to the RC and note the signal strength. Many lost-link problems can be avoided just by paying attention to signal strength and not flying too far away. Also, keep in mind that you're required to maintain visual line of sight (VLOS) with your drone. Many drones support transmission ranges far beyond your line of sight, but just because your drone can fly far away from you doesn't make it legal.

>> **Poor weather conditions:** If you must fly in poor weather conditions, remember that heavy cloud coverage and static electricity in the air can impact the link between the drone and RC. In other words, if you must fly in

these conditions, you should greatly raise your awareness level. See Chapter 14 for a deeper dive into clouds and weather conditions.

>> **Obstacles:** If you need to fly around tall buildings and other structures, keep in mind that these obstacles can interfere with your signal. For many drone piloting jobs, you may need to fly around structures like these, but just be aware that some structures interfere with radio signals and can cause link problems.

>> **Software problems:** Drone software is made by people, so it's not perfect. A software glitch or failure can cause a lost link. Be proactive: Always install updates in a timely fashion. These updates often fix known problems, so be sure to keep the software updated.

>> **Interference:** Finally, electrical interference, power lines, power plants, other radio signals, and different kinds of electromagnetic interference can all cause link problems. You can't always predict the presence of this kind of interference, but just know that there is always this "invisible" element when you fly.

Dealing with a lost link

Fortunately, you have several strategies to try that can help you recover the link and get your drone back safely, as the following sections describe.

Use Return to Home

Return to Home (RTH) is a mission-critical safety feature of modern drones that your drone may have. Unfortunately, people often treat RTH as a novice feature, thinking that if they can fly the drone, they can return and land without using the automated RTH features. That's great — as long as you have a link. If you don't have a link, the drone is designed to invoke RTH and come home by itself, which is exactly what you want. So, RTH is an emergency response strategy that you must set up *before* you experience a lost link.

You can configure the RTH feature in your RC's menu system. For most drones, you select an option for RTH to take over if the link is lost. You can determine the altitude at which to fly in order to return to home, as shown in Figure 15-3. Some drones give you other options, such as hovering or descending, but generally RTH is better (see Figure 15-4).

Keep in mind that when you power up the drone and RC, you'll need a few minutes to acquire a GPS signal so that the drone can set a home point. Without the home point, the drone won't know where home is and RTH won't help you. Sometimes drone pilots get in a hurry to take off. Don't be in a hurry; instead, give the drone time to set the home point. In the case of an emergency, you'll be glad you did.

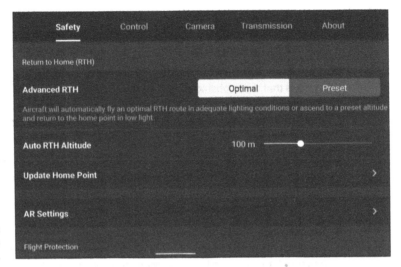

FIGURE 15-3:
Configure RTH on
the controller.

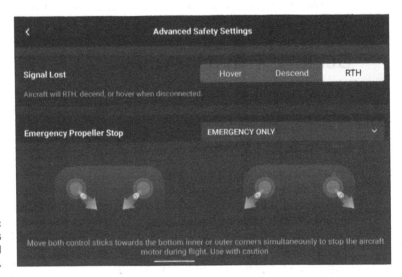

FIGURE 15-4:
Some drones
have additional
lost-link options.

REMEMBER

Although RTH is a critical feature, it also isn't fail proof. RTH depends on GPS, so if the drone loses the GPS signal, it has no way of knowing where the home point is. For this reason and many others, the FAA always recommends that you maintain a visual line of sight. Read on!

Maintain visual line of sight

In Chapter 6, I mention that maintaining visual line of sight (VLOS) is an FAA requirement for all drone pilots. In fact, I mention it often in this book in various ways because at the end of the day, VLOS can help you avoid all kinds of problems. One of those is the case of a lost link.

Sometimes lost links affect only the downlink (which transmits information from the drone to the RC). Without the downlink, you may be able to still fly the drone using the uplink (depending on the drone model), but you won't have any live video on the screen. If you can physically see your drone in the sky, you can fly it back toward you in hopes of regaining the link, or at least you can manually land it using the RC. Of course, if you can't see the drone in the sky, you have no idea where the drone is and how to fly it back toward you. This is a practical application of VLOS: In the case of a lost downlink, VLOS can save the day.

Restart the RC

If you lose the link and can't seem to get it back, you may be able to restart the RC and get the link reestablished. The success of this strategy depends on what caused the lost link in the first place, and whether the drone has already initiated RTH, but in some cases, an RC restart may solve the problem. Just keep this in mind.

Have pre-planned landing points

If you're flying a complex mission during which link loss could be caused by interference and obstructions, it's a good idea to scout the location and establish several possible landing points, or what the FAA calls pre-planned *flight termination points (FTP)*. The idea is that if something is going wrong with the link, or maybe you keep losing and reestablishing the link, you can land in multiple places so that you can troubleshoot.

If you lose the drone, try to find it

In some cases, you simply lose the link and the drone flies away because of system problems or GPS loss. In this case, note on your controller the last known location, and then try to find and recover the drone. If you have insurance, most policies require you to try to recover it. However, do so only if you can do it safely. Now is not the time to dive into the ocean looking for your drone.

TIP

Also, you may need to report a lost link and lost drone, especially if you're flying near an airport. In that case, it's best to be proactive and notify the ATC for that airport. Also, if you're flying in an urban area, notify local law enforcement. A flyaway drone is going to crash eventually, so letting people know is always a good move.

AT THE RISK OF SOUNDING LIKE A BROKEN RECORD . . .

It amazes me how often people pay good money for a drone with safety features and then don't use them. If your drone has sensors, put them to use! Configure the RC for braking or bypass if an obstacle is detected, and always use RTH. These simple features that you've already paid for can help you avoid an in-flight emergency.

Also, never forget about batteries. Make sure the drone batteries are charged before you fly. If you notice that a battery doesn't charge fully or doesn't seem to last as long as it should, buy a new one. Yes, they're expensive, but you don't want a bad battery causing your drone to fall out of the sky. Play it safe: Never use a questionable battery!

Chapter **16**

Becoming a Safe, Responsible Drone Pilot

A chapter with a title about becoming a safe drone pilot may be tempting to skip. You can think, "I'm safe and responsible — moving on!" But not so fast. The topic of "safe and responsible" drone piloting includes some technical information about lift, stalls, loading, and such. The FAA expects you to understand this information, so much so that if you take the Part 107 exam, I bet you'll see a question or two about it. Some of the questions about loading concern fixed-wing drones. Although you probably won't fly a fixed-wing drone unless you're entering a specialized field of drone piloting, the Part 107 exam expects you to know a couple of facts about them. I mention those details in this chapter but tell you only what you need to know.

Beyond the technical information you gain in this chapter, the information here can help you in a variety of missions, especially as your piloting skills develop and you start performing professional duties. In fact, depending on what kind of drone piloting you plan to do, you may use some of this information repeatedly. In this chapter, you delve into all the factors that you must take into account when you add any weight-bearing item to a drone. You also discover several approaches to making a good judgment call in a pinch, as well as identify various attitudes that can get you in trouble when you fly.

Understanding Loading

When you think of a drone, you probably think of the model you currently own, and you likely focus on how you use that drone. However, in the professional sphere of drone piloting, there are all kinds of drones that are used for a variety of purposes, not just photography.

For example, agricultural drones are designed to carry fertilizer, pesticides, and other chemicals for distribution on crops. Also, some drones that are specifically designed for inspections can carry different kinds of inspection equipment.

Of course, anything you add to a drone (called "loading") changes the weight the drone must manage as it flies.

Every drone model is a bit different, and drone manufacturers publish information about weight loading and distribution for a particular model. Although this chapter's reading may not be the most exciting, you need to understand not only you're drone's features (what it can do) but also its limitations (what it can't do).

The first key to understanding your drone's loading capabilities is to look at the manufacturer's published data about them.

Force equilibrium

When you fly your drone, you put the following four forces on it (see Figure 16-1):

>> **Thrust:** The power that pushes the drone in the particular direction you want. On a quadcopter, all the propellers and the motor work together to give you thrust so that you can move the aircraft forward, backward, or from side to side. (Chapter 1 provides details about how the propellers work.)

>> **Drag:** The force that works against thrust. You can think of drag as the friction of the air molecules against the drone as well as the pressure of that air. Different atmospheric conditions create more or less drag, so your drone has to work harder or less hard to create thrust, depending on the air.

>> **Lift:** The force that enables your drone to ascend. Differences in air density and pressure affect lift, as does the weight of the drone.

>> **Weight:** The opposite force to lift. Weight is effectively the force of gravity that pulls everything back to earth.

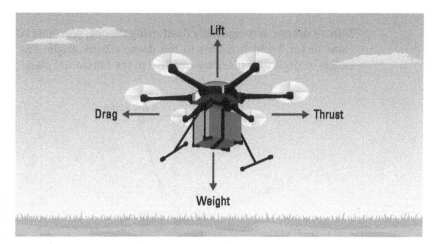

FIGURE 16-1:
The four forces
enacted
on a drone.

When a drone is straight, level, and hovering, thrust is equal to drag and lift is equal to weight. In other words, you have a force *equilibrium*. When you add motion or additional weight to the drone, that force impacts force equilibrium, which is why understanding the four forces matters. When you change loading on a drone, that change affects the drone's ability to fly or maintain flight.

REMEMBER

You need to consider another aspect of weight — and balancing that weight — when loading your drone. Your drone manufacturer will tell you about loading and balancing and what additional weight the drone can safely manage with four forces in play. However, it's important to realize that the maximum weight is figured based on optimal flying conditions, which, of course, you don't always have!

For example, say you need to fly a loaded drone in Denver. At mile-high elevation, the air is thinner, which means that to take off from Denver, the drone's motors will have to work much harder to achieve lift than it would if taking off from Houston. As another example, flying into the wind on a rather windy day creates more drag. So just because the drone manufacturer tells you that a drone can support a particular weight, you can't necessarily fly with that maximum weight limit every time. The flight conditions affect the drone's ability, so pilots have to think carefully when loading a drone near its maximum weight. Unfortunately, no simple math formula exists to follow in these cases. This is where pilot judgment comes into play.

Figuring load factor

As you think about drone piloting, you're probably thinking of quadcopters. However, not all drones are quadcopters. Some are fixed-wing drones, and your drone piloting career could lead you to piloting one of these drone types.

With all drones, but especially fixed-wing drones, you must take into account the "load factor," which relates to the drone's bank angle. The bank angle is the incline or decline of the wings relative to the horizontal plane (see Figure 16-2).

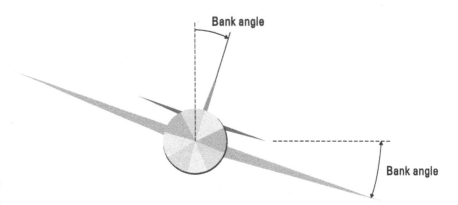

FIGURE 16-2:
Bank angle.

Bank angles are created by turning the drone, and the bank angle puts more force on the wings of the aircraft. More force affects loading and how much load a drone can manage. Bank angles are calculated in degrees, and the FAA provides a chart to help you figure the load that will be placed on the drone relative to the bank angle. If you're taking the Part 107 exam, don't be surprised if you get a question to this effect.

So, how do you calculate the load factor? The chart you need (see Figure 16-3) is in the FAA's testing supplement, which you can download at https://www.faa.gov/training_testing/testing/supplements/media/sport_rec_private_akts.pdf.

If you increase your bank angle as you fly, such as by making a turn, that action also increases the load factor. The greater the bank angle, the higher the load factor. In the end, the load factor is a measurement of what the drone's wings can support divided by the actual weight placed on them during banking. In mathematical terms, you multiply the load factor found on the FAA's chart by the actual weight of the drone and its contents. That equation tells you the total load that has to be supported by the aircraft's wings.

REMEMBER

Here's an example using the chart in Figure 16-3. Say you have a drone that weighs 20 pounds, including the extra weight you need to add to it for a mission. In that mission, you need to make a 30-degree bank angle. If you look beside 30 degrees on the chart, you see that the load factor is 1.154. Multiplying 1.154 by 20 pounds comes to 23.08 pounds. The drone's wings must be able to support 23.08 pounds to make a 30-degree bank angle safely.

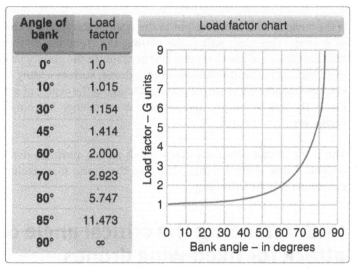

Angle of bank φ	Load factor n
0°	1.0
10°	1.015
30°	1.154
45°	1.414
60°	2.000
70°	2.923
80°	5.747
85°	11.473
90°	∞

FIGURE 16-3: The FAA's load factor chart.

Source: FAA-CT-8080-2H Airman Knowledge Testing Supplement for Sport Pilot, Recreational Pilot, Remote Pilot, and Private Pilot. https://www.faa.gov/training_testing/testing/supplements/media/sport_rec_private_akts.pdf

Here's one more example. Say that a drone weighing a total of 48 pounds needs to make a 70-degree bank angle. What's the load factor? Looking at the chart, you see the number 2.923 beside 70 degrees, so you multiply 2.923 by 48 pounds, which equals 140.3. Rounding up, you know that this drone must be able to support 141 pounds. As you can see, bank angles can drastically add to the weight of the drone!

Keeping in mind the big picture, drone pilots use this kind of information to determine how much weight a drone can carry and the maximum bank angles that can be made in order to support that weight. Very important!

Taking center of gravity into account

Along with the concepts of loading and weight, you also have to consider the balance issue, or center of gravity. Drones are engineered so keep weight balanced. However, when you add weight to a drone, that additional weight needs to be balanced as well. That's why studying the manufacturer's guidelines and instructions is very important when you add payload. You want the center of gravity in the center of the drone, not to the front, back, or one side. That weight needs to be balanced for the drone to fly well.

TIP

If the payload adds more weight to the front of the drone, it throws off the center of gravity. This shift will make getting the right lift or thrust more difficult because the drone has to try and compensate for the gravity problem on the front. Even making hardware changes to your drone, such as adding or changing a camera, can impact this center of gravity. Changing the center of gravity can make the drone fly less smoothly and require more energy to complete normal flight maneuvers.

The point is that when you think about adding payload, don't just think about weight. Think about balance also. Use the drone manufacturer's guidelines to determine weight balancing.

Understanding the critical angle of attack for fixed-wing drones

Another issue to note for fixed-wing drones is the concept of critical angle of attack. This concept relates to stalls, which isn't something you have to worry much about with quadcopter-type drones. But for fixed-wing drones, it's important to understand.

First, a stall happens when the airflow over the wings is disrupted. This disruption can cause a sudden loss of lift, which pitches the nose of the drone down. As you can imagine, a drone flying downward isn't what you want because a sudden loss of altitude can lead to a crash. The angle of attack is the angle from the front of a wing to the back. The angle has to keep air moving over the wings in order to create lift. You need lift if the drone is climbing, descending, or even flying in a level manner. No matter how the drone is flying, lift is necessary to counteract the drone's weight. The *critical* angle of attack is the point where the angle of attack reaches its maximum lift.

Stalls typically occur when the critical angle of attack is exceeded. Here's what that means: Wind passes over the wings as a fixed-wing drone flies forward. The angle of the wings affects air flow. So, the critical angle of attack is the angle of the wings in relation to air flow. A drone has a maximum angle of attack depending on how the drone is manufactured. When a pilot exceeds this angle of attack, the airflow gets disrupted as it passes over the wings (too much angle), which causes a sudden and dramatic loss of lift, as shown in Figure 16-4.

TIP

I've included this information here because it goes along with weight and load balancing (center of gravity). All these factors can affect the critical angle of attack and what angle can produce a stall for a fixed-wing drone. Again, the key is to carefully study the manufacturer's guidelines and data for weight, balancing, and center of gravity so that you can understand what a drone can and cannot do as you think about weight, balancing, and the atmospheric conditions as you fly.

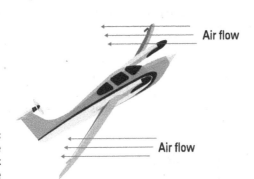

Critical angle of attack exceeded

Air flow

Air flow

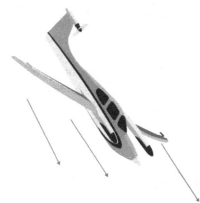

Loss of lift result in a stall

FIGURE 16-4:
Exceeding the angle of attack can produce a stall.

Making Sound Flight Judgment Calls

The FAA provides a framework that can help pilots make sound judgment calls before and during flight. This framework, called Aeronautical Decision Making (ADM), is effectively a model that helps you think in the right direction. The framework is designed to be continuously at work in a pilot's mind so that the pilot can make the best flight decisions based on situations as they occur. You can find out more about ADM at https://www.faa.gov/newsroom/safety-briefing/aeronautical-decision-making-adm, but a quick overview follows.

ADM uses a 3-P model: Perceive, Process, and Perform:

» **Perceive:** Gather all information and think through the flight mission. This step occurs before the flight but also continues during the flight based on changes that may occur (such as weather conditions, obstructions, and even battery power). The point is that you, as the pilot, fully perceive the details of a mission from start to finish, but you continue perceiving that information as details change during the mission.

» **Process:** As you perceive, or understand, the elements of a mission, you mentally process that information and consider its impact on flight safety as well as the completion of the mission. The act of processing information continues as you fly.

» **Perform:** You implement the best course of action. After gathering information, you decide and execute the best, safest action for your mission.

REMEMBER

In practical terms, pilots should go through a mental process to make safe decisions. For example, say you need to perform a mission in a highly populated area that involves several technical drone moves. From a safety point of view, you may need to consider altering the parameters of the mission or even moving the timetable of the mission to take place when fewer people are in the area. You can use this same process when you perceive other potential risks, such as weather. Obviously, you can't avoid every possible risk, but the goal is to assess every mission and make decisions that lead to risk mitigation as much as possible.

Considering Other Models

Along with ADM, the FAA provides a few other models that drone pilots should consider using as a mental framework to maintain while flying. These are the DECIDE, IMSAFE, and PAVE models, and the following sections look at each of them.

DECIDE model

The DECIDE model is a framework to help pilots make decisions. It works under the category of ADM, but you may find that the DECIDE model is a bit more intuitive and easier to think through on the fly (so to speak). The DECIDE model really comes into play when something changes in a mission, and it can be a helpful thinking model when you need to make a flight decision.

DECIDE is an acronym that means the following:

» **Detect the problem.** The first step is an element of situational awareness that involves detecting that a problem exists and needs to be addressed. For example, say you're in the middle of a mission and suddenly you notice a bank of dark clouds forming. The DECIDE model of thinking wants you to immediately detect this as a possible threat and not ignore it.

» **Estimate the need to react.** Is the problem significant enough that you need to react at the moment? Do you need to change something in the flight? Do you need to land and reassess the situation? These questions relate to the pilot's ability make sound judgment calls as to whether the mission can continue safely or if a landing is needed.

» **Choose an action.** Based on your estimation, what is the best course of action to take? Also, are other courses of action available that you didn't previously plan?

- » **Identify a solution.** After you consider the possible actions, you choose specific actions or substeps of those actions. For example, if the best solution is to land as quickly as possible, what steps will get your drone safely on the ground in the quickest manner?

- » **Do the necessary steps.** At this point in the DECIDE model, you perform the steps you've chosen. You implement whatever steps are needed, but you also realize that you may need to modify those steps as you are taking them.

- » **Evaluate your action.** After taking the necessary action steps, consider your action. Was the action effective? Did it mitigate the risk? Would another action be better next time?

REMEMBER

It's easy to read through a model like this and sort of ignore it, but remember that the DECIDE model is a way to train your mind to think. This model can help you make decisions quickly when something changes during a mission, and in most cases, it can help you make the best, safest decision available to you. Using the model can help you make those good decisions much more quickly because you've trained your brain to follow a framework.

IMSAFE model

Along with making sound flight decisions, you also need to take your own condition, as the pilot, into account. The fact is, most aviation accidents occur because of human — not mechanical — error. For this reason, the FAA wants drone and crewed-aircraft pilots to seriously assess themselves. The IMSAFE model, described in the following list, is designed to help you do that:

- » **Illness.** Think about your general health. Have you been or are you currently sick? Have you been sick, and are not fully recovered? Even a bad cold can impact your thinking ability and reaction time. So, seriously ask yourself, "Am I well enough to fly?"

- » **Medication.** Medications can have a serious impact on your reaction time. Are you taking any medication that could impair your ability to make decisions? Have you taken any medications that could make you nervous, jumpy, or drowsy, causing you to react too quickly or slowly? If you're unsure, you can contact an aviation medical examiner (AME) and ask questions about various medications. You can find out more information about AMEs in your area at https://www.faa.gov/pilots/amelocator.

- » **Stress.** Am I under too much stress to fly safely? Consider not only the mission but also personal stress. Is there a stressful situation in your life that is dominating your thoughts? If so, you may not be ready to fly.

- **Alcohol.** Even a small amount of alcohol impairs judgment. Follow the FAA's regulations concerning flights and alcohol use. See Chapter 7 for details.

- **Fatigue.** A lack of rest has a direct impact on reaction time, focus, and overall health. Make sure you've had plenty of sleep before you fly.

- **Emotions.** How's your general emotional state? If you're dealing with anxiety or emotional distress, your situational awareness can be greatly impacted.

PAVE model

One final model for you to consider essentially merges the DECIDE and IMSAFE models into a more global model called PAVE:

- **Pilot in Command (PIC):** The pilot considers themselves before the flight and during the flight. Are you ready to fly safely? Consider the IMSAFE model to evaluate yourself before you fly.

- **Aircraft:** Is the aircraft ready to fly? Have you performed all pre-flight safety checks? Are the batteries charged? Is the aircraft capable of performing the intended mission? Do any updates, maintenance, or modifications need to be performed?

- **Environment:** Have you taken a careful look at the forecast? Have you considered other environmental variables such as density altitude (see Chapter 14)? Are there possible terrain issues or obstructions you should consider? (And in case you're wondering, the *v* in the PAVE acronym comes from the *v* in *Environment*.)

- **External pressure:** Stress, time constraints, and distractions can all impact the safety of your mission. Have you considered all external pressures that may impact your flight?

Getting Rid of Incorrect Thinking

One of the problems the FAA models address concerns pilots and incorrect thinking. There are really attitudes, or ways of thinking that endanger drone missions, property, and potentially people. I'm sorry to say that these incorrect ways of thinking plague drone user groups on social media I see all the time. They're dangerous, and they cause problems. Remember, most drone accidents happen because of pilots, not drone hardware.

Yet, everyone is susceptible to wrong thinking and attitudes, so it's a good idea to think through the following sections and consider whether you struggle with any of the types of thinking they go over. Be honest and consider the right way to think. You can work to remove wrong attitudes from your mind and adopt correct, critical thinking. Professional pilots do, and if you want to be a successful professional drone pilot, the right attitudes will serve you well, so keep reading.

Anti-authority

An anti-authority attitude leads to ignoring drone piloting best practices and even FAA rules, such as flying in the correct airspace (or getting permission to fly in an airspace when you need to). Sometimes the anti-authority attitude is more subtle. It's a general method of thinking that says, "I know the rules, but they're not that important in my case." Although drone piloting rules and regulations can be a bit overwhelming at times, those rules exist because studies prove that the rules keep property and people safe. And in the end, that's most important. You never want to find yourself in a situation where your piloting behavior has harmed another person.

TIP

So think carefully: Do you have at least a bit of an anti-authority attitude? If you can honestly answer "yes," you've just won half the battle. At this point, you can simply start repeating to yourself, "The rules exist for a reason. I want to be a safe and effective pilot. I never want to hurt someone." Filling your mind consistently with these statements can help you eject anti-authority attitudes from your thinking.

Impulsivity

By definition, impulsive means to act without forethought. This attitude can be devastating for drone pilots. In fact, a lot of this chapter has explored models designed to help you think before you act, and the ability to think before you act when you fly has a lot to do with safety and success.

An impulsive pilot tends to act without considering the possible consequences. Just think, would you really want to be on a commercial flight with a pilot who was impulsive? That thought is a bit horrifying, so consider that idea with drone piloting as well.

Some of us tend to be more impulsive than others; often we're just born that way. So if you know you tend to be a bit impulsive, you just have to do the work of ejecting that thinking from your mind. After all, impulsive thinking is what leads to impulsive actions.

TIP

If you tend to be a bit impulsive, just work on mentally slowing down. In many piloting situations, the first response is to maintain control of your drone. After that, you can often stop, hover, and decide what to do next. Unless the weather is creating an imminent problem and you need to land quickly, just hover and think for a moment before you make a decision. As you think, ask yourself, "What are my options?" You usually have more than one, so consider them before you decide what to do next.

Invulnerability

An invulnerable attitude means that you just don't believe something bad will happen to you. You can read about all kinds of drone piloting errors, accidents, and crashes, and just think, "That won't happen to me."

Not so fast! Think about a drone pilot who crashes their drone into a car. Guess what? That's not the first pilot who has crashed into a car. In fact, you can think of almost any accident or problem and realize those kinds of issues happen over and over.

So why not to you? You're just a person like anyone else, and common problems can certainly happen to you or me. Therefore, the correct thinking in any situation is simply, "That could happen to me!" If you hear a story of a drone pilot who made a terrible mistake, keep in mind that we're all susceptible to making terrible mistakes and work on ejecting this thinking from your mind. Instead, think, "Anything bad can happen to me." When you think that way, you'll make better decisions and fly with a greater sense of caution.

Machoism

Another attitude that the FAA identifies as a problem is machoism, which is an attitude of overconfidence and a need to impress others or to prove oneself. Macho pilots are often tempted to engage in risky flight behavior, ignore rules and best practices, and fly beyond their current skill set.

Machoism can lurk under the surface of a person's consciousness. It's a general feeling that you need to impress others or show a consistent level of confidence or control. Yet, a professional drone pilot should be quick to admit that piloting comes with certain risks that are outside your control. You can't control the weather, for example!

Work on developing an attitude of safety. Work on ejecting the need to impress other people with your drone piloting. Getting rid of these attitudes will help you fly better in the long run, and certainly with safety.

Resignation

The opposite attitude of machoism is resignation. This attitude belongs to pilots who feel easily defeated or helpless, especially in the face of difficult piloting circumstances. This attitude can cause passive behavior instead of active behavior, which can lead to slow responses and safety concerns.

Along with the attitude of resignation is a general feeling of compliance. A pilot with resignation problems may agree to attempt a mission that isn't safe, feeling that they shouldn't say "no" to requests from clients.

In the end, you have to say to yourself, "I am the remote pilot in command. I can make good decisions. I can execute safe and effective actions." You can't control everything, but you can control a lot. Putting effective piloting skills to work goes a long way toward safety and success. Be the pilot!

KNOW WHEN TO ABORT

Sometimes a pilot thinks, "I'm a good drone pilot and I can handle this situation." That's a good thought, but during a flight, things may not go the way you want them to. Maybe the weather isn't great and suddenly it's too windy. Or, in your pre-flight planning, you didn't know about some of the obstacles you're facing, such as guide wires (which are wires that support tall, slender objects, such as radio towers) and powerlines. In those cases, the smartest things you can do is land and reassess.

However, the issue of pride may enter in. Some drone pilots really struggle with feeling like a failure if they have to land and not complete a mission. Fight back against this feeling! To reiterate, drones are machines, and sometimes things go wrong. But in most cases, crashes happen because pilots are either careless or just make poor piloting decisions. No pilot is perfect, but a safe pilot will be quick to abort a mission and reconsider the mission parameters. You may just need to make a few modifications to your plan before you fly again. The safe thing to do when you're unsure is abort the mission and land. This is not a sign of failure — it's a sign of a smart, effective drone pilot.

5

Getting to Work: Drone Piloting in the Marketplace

Chapter **17**

Taking Great Drone Photos

A drone is basically a flying camera. Think about it: The camera allows you to see where you're flying, but you can also use the camera to take photos and videos. In almost all drone piloting careers, that's what you'll spend your time doing. After all, the point of flying a drone is to help you and your client see what's above your field of vision. That's where taking good photos comes into play.

Taking photos with your drone is much different from taking photos from the ground. In the sky, the angles you're shooting are different, the lighting is different, the objects on the ground may be reflective, and generally speaking, drone shots tend to be overexposed.

Fortunately, after you understand the challenges to getting great drone photos, you can employ some simple strategies to help you overcome those challenges. This chapter walks you through photography techniques, file types and resolution issues, the use of good angles, and basic photo editing. Think of this chapter as a crash course in taking great drone photos. And remember, practice makes perfect, so be sure to apply what you read in this chapter and practice taking great shots.

TIP

Entire books are written about photography and exposure, meaning that there is a lot to learn. However, in this chapter, you'll find all the basics specific to drone photography and can be taking great photos in no time, but if you want more information, you may want to pick up a general-photography book or check out some good photography videos on YouTube.com (www.youtube.com). Here's a good one to get you started: https://www.youtube.com/watch?v=200eLwrN4jo.

Considering the Basics of Good Drone Photography

Good photography always comes down to two fundamental components: hardware and skill. A digital camera is a machine that performs all kinds of electronic functions, but the lens quality of that camera has everything to with photo quality, and good lenses are expensive. A simple point-and-shoot camera just can't shoot the same photo quality as a good DSLR camera. The lenses you might use on a DSLR may cost much more than the entire point-and-shoot camera.

This principle is true for drones as well. The price point of the drone you're flying has a lot to do with its camera quality. Drone cameras are expensive to make because they have to be small and lightweight. Basically, drone manufacturers try to give you the power of a DSLR in a very small package. For example, the DJI Mini series has a good camera and does a great job. Yet, the DJI Mavic series cameras are much better, and those models take better photos. Those models are also much more expensive.

The point is that there is always a balancing act between getting the best drone with the best camera for your shooting needs and having the finances to do so. In the photography world, professionals are always investing more money to try to get the best lenses for the best shots. You'll have to juggle that balancing act as well. Try to get the best drone with the best camera you can afford. No amount of photography skill will overcome poor equipment.

But the reverse can also be true. Good equipment doesn't necessarily equal good photos. You need to know a thing or two about drone photography so that you can use your drone to get the best shots.

Understanding shooting modes

Your drone probably provides a few different shooting modes. A *shooting mode* determines how the camera takes a photo and how much you get to directly interact with the exposure settings for that shot. Following are a few common shooting modes. Your drone may have only a few of these or all of them, depending on the drone model:

>> **Auto:** All drones with a camera have an auto mode. Auto mode means *automatic,* and with this mode, the drone camera makes all the exposure decisions. When you get ready to take a photo, the camera takes a light-meter reading and decides the exposure settings for that particular shot as well as the focusing setting. With modern camera equipment, your drone's auto mode will typically do a good job of choosing those settings, and Auto mode is a quick and easy way to capture good photos — most of the time.

>> **Pro or Manual:** Your drone may have a Pro or Manual mode. With this mode, you get to make all the exposure decisions for the camera, such as shutter speed, aperture, and ISO. You also choose focusing settings in this mode. For example, in Figure 17-1, I have Pro mode selected and am currently choosing a shutter speed for the shot. You choose these settings on the controller as your drone is in flight.

Shutter speed adjustment

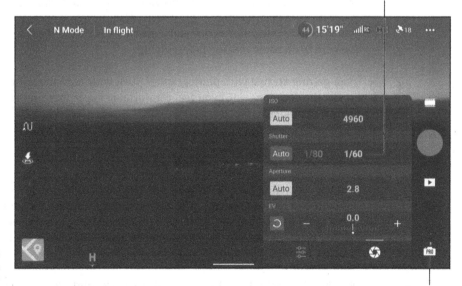

FIGURE 17-1:
Pro mode
showing shutter
speed selection.

Pro mode

- >> **Automatic Exposure (AE):** Your drone may have an AE mode, which means the drone camera chooses exposure settings based on other choices you make. This mode works differently on different drones, but generally, this mode allows you to choose a setting (such as a faster shutter speed), and the drone will automatically choose other settings to try to give you the best exposure.

- >> **Manual Focusing (MF):** Your drone may provide a manual focusing mode, giving you the power to override automatic focusing so that you can get the right focusing for a shot you want. This mode is helpful in tricky focusing situations.

- >> **Burst:** Burst mode allows you to take a bunch of photos in rapid succession. This mode is helpful with action shots, such as wildlife on the move or even waterfalls. The idea is that you can take a bunch of photos of a moving target very quickly to try to get one or two you really like.

- >> **Panorama:** Panorama mode is a type of automatic mode that enables the drone camera to take several photos of a scene and then automatically merges those shots to create a panoramic scene. This can be a great mode to use for large-scale landscape shots, such as a scene from a mountain top.

- >> **Timelapse:** In Timelapse mode, the camera automatically takes a series of images of a period of time. For example, you can use Timelapse to take a photo every ten seconds of a sunset, giving you a collection of interval shots.

As a reminder, different drones provide different modes, so be sure to check out your drone's documentation for details. Also, how you access different modes on your drone's controller varies and sometimes isn't the most intuitive process, so be sure to do some reading before flying.

TIP

Just getting started with photography? Stick with Auto mode first. Let your drone make the decisions until you feel comfortable experimenting with more complex shooting modes.

Focusing on Exposure

Camera exposure is a collection of settings that attempt to give you the right mix of light and color in a photo. The three primary settings are shutter speed, aperture, and ISO. In Auto mode, the camera makes the exposure settings for you, but if you want to use Pro or Manual mode, you adjust each of these settings. Of course, to adjust them, you need to understand them.

It's helpful to think of shutter speed, aperture, and ISO as an exposure triangle, depicted in Figure 17-2, with the perfect photo coming together in the center of that triangle. Each of these settings plays a big part in the exposure and, therefore, the quality of the photo you shoot.

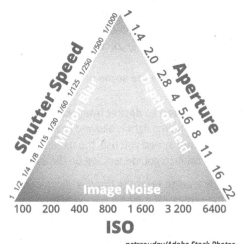

The Exposure Triangle

FIGURE 17-2:
The exposure
triangle.

petrroudny/Adobe Stock Photos

Shutter speed

When you take a photo, the camera's shutter opens and closes, exposing the sensor to light. *Shutter speed* refers to the time the shutter is allowed to stay open. A slow shutter speed keeps the shutter open longer, whereas a faster shutter speed opens and closes the shutter faster. The shutter speed affects exposure because the amount of time the light is allowed to act on the sensor determines the darkness and brightness of a shot.

Also, shutter speed can blur or freeze action. For example, in Figure 17-3, the slow shutter speed blurs the motion of the waterfall on the left, whereas the fast shutter speed freezes the action of the waterfall on the right.

FIGURE 17-3:
Slow shutter
speeds blur
action (left image),
and faster shutter
speed freezes
action (right).

davelogan/Getty Images

Lorado/Getty Images

Shutter speed is expressed in fractions of seconds. For example, if you see a shutter speed notation of 1/25, that means the shutter was open for 1/25th of a second. Sometimes you'll see shutter speeds expressed in whole numbers, such as 25, 50, 200, and so forth, but these mean 1/25, 1/50, and 1/200. Just keep that in mind.

In terms of using shutter speed, here are some quick tips:

>> **Faster shutter speeds can lead to darker images.** Remember, the shutter speed determines how much time light is allowed to act on the camera sensor. The faster the shutter speed you use, the darker (meaning underexposed) the image you're likely to get, depending on the available light.

>> **Slower shutter speeds lead to brighter images.** Slower shutter speeds allow light to act on the sensor for a longer period of time, often leading to brighter (overexposed) images, depending on available light.

>> **Distance affects freezing action.** The closer your drone is to a moving object (or the closer you're zoomed in), the faster the shutter speed you will need to freeze the action. For example, if you're photographing a herd of deer running in the distance, 1/500 is probably enough to freeze the action of the deer. If you're closer to the deer, you may need a shutter speed of 1/1000 or more to freeze the action.

>> **Slower shutter speeds can help you get better image color in lower light.** If you're shooting in lower light conditions, such as at dusk or on a very cloudy day, a slower shutter speed can help you get more light in your image, and therefore more color. However, finding the sweet spot means that you often have to experiment with various shutter speeds. Note that slow shutter speeds can cause image problems because a slight drone movement (such as wind) can cause the image to come out blurry.

TIP

The shutter speed range available to you depends on your drone's camera. Some drones may support shutter speeds as low as one second while also supporting several thousandths of a second at the top of the range. The better the drone camera, the better the range.

Aperture

As noted in the previous section, shutter speed refers to the amount of time that light is allowed to act on the camera sensor. *Aperture*, on the other hand, refers to the actual amount of light let in to the camera's sensor. Your drone's camera lens has aperture blades inside that open and close to allow less or more light to enter, depending on the aperture setting the drone selects in Auto mode, or the setting you select in Pro or Manual mode.

Aperture is expressed with an *f* followed by a number, such as f/4 or f/22. Aperture is often confusing because small *f* numbers refer to aperture blades that are more open, whereas bigger aperture numbers refer to blades that are more closed. So, f/2 is considered a "big" aperture, whereas f/22 is considered a "small" aperture. Confusing, huh? Small numbers mean big and big numbers mean small.

Here's how to cut through the confusion: The aperture numbers refer to the diameter opening of the aperture blades (see Figure 17-4). Think of those numbers as fractions. 1/2 is much bigger than 1/22. If you have a water hose that's a 1/2" diameter and one that is 1/22" diameter, the 1/2" hose allows more water to flow through it. It's "bigger." Aperture works the same way. An f/2 aperture allows much more light to hit the camera sensor than an aperture of f/22.

FIGURE 17-4:
Small aperture numbers mean more light, whereas big aperture numbers mean less light.

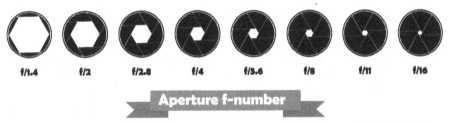

f/1.4   f/2   f/2.8   f/4   f/5.6   f/8   f/11   f/16

Aperture f-number

There is a lot to learn about using aperture, but here are the main things you need to know:

>> **Small aperture numbers mean more light.** Small aperture numbers (which equal a big aperture), such as f/2, mean that more light hits the camera sensor. This can be great when you're shooting in low light conditions but will also overexpose your photo if plenty of light is available.

>> **Big aperture numbers mean less light.** Big aperture numbers (which equal a small aperture), such as f/22, mean that less light hits the camera sensor. Using a big aperture can be helpful on bright days, but if you don't have enough available light, you'll often get underexposed photos.

>> **Small aperture numbers can create a shallow depth of field.** Depth of field refers to the areas of a photo that are acceptably sharp. Because of light refraction inside the lens, big apertures (small numbers), such as f/2, can blur the background of an image when you're focused on an object that is close to you. This effect may or may or not be desirable, depending on what you're shooting.

>> **Big aperture numbers can create a deep depth of field.** When you use a big aperture number, the aperture blades have a small opening, which means less light refraction. A bigger aperture number, such as f/22, can give you a deep depth of field, meaning that most of the photo will stay sharp and in focus.

ISO

ISO stands for International Organization for Standardization (which is a mouthful). In short, ISO is a carry-over term from old film days. ISO referred to the sensitivity of a particular film to light. Today, we use the same term to refer to the sensitivity to light of a camera sensor rather than film. The good news is that ISO is easier to understand than shutter speed and aperture. Most cameras use a default and lowest ISO setting of 100, and then the numbers go up from there. The higher the number, the more sensitive the sensor becomes to light.

Here's the deal: Higher ISO numbers can help you capture a good image when lighting is low. Otherwise, just leave the setting at its lowest value. The problem with ISO is that higher settings also tend to introduce more graininess, so the image quality tends to go down.

TIP

For this reason, most photographers tend to make adjustments to shutter speed and aperture to get the right exposure while leaving ISO at its lowest value. Sure, there are times when you need to turn up the sensitivity if lighting conditions are poor, but always start with shutter speed and aperture adjustments first.

Considering File Types and Resolution

When you take a photo with your drone, you create an electronic photo. That photo will be a certain kind of file type. These file types conform to certain industry standards for how the photo data is recorded, compressed, and formatted. A file type enables different kinds of programs to open and display a photo. At the same time, you'll shoot photos in a certain resolution, which impacts the quality of the photo as well as the file size of the photo (in megabytes).

Drone file types

Most drones support a few different file types. Your drone may support all three of the following, or at least two:

>> **JPEG:** Virtually every camera supports JPEG images, including drones, point-and-shoot cameras, and phones. The internet is made of JPEG images, and for good reason. The JPEG file format produces good images that have low file sizes (fewer megabytes), so they're easy to use, easy to upload and display, and easy to download. The default file format that your drone uses is most likely JPEG. If you're shooting photos for a client, that client will most likely want JPEG images so that the photos can be used in a variety of ways.

However, know this: JPEG images are compressed and are considered "lossy." *Lossy* means that if you do heavy editing to a JPEG image, information keeps getting discarded during the editing process, which can lead to poor images. Just remember that JPEG is an industry standard, but it's not the best file format to use if you plan to do a lot of editing.

>> **PNG:** PNG is another industry standard file format that most programs can read and display. PNG files are also used frequently on internet web pages. PNG files are great, and they're not as lossy as JPEG images. However, the file size (in megabytes) may be significantly larger than a JPEG image. That's fine as long as you have plenty of storage space on your drone's memory card and your client is fine with bigger file sizes.

>> **RAW or DNG:** Your drone may support RAW files or DNG files. You can think of both of these as a *digital negative*. RAW and DNG files have no compression and contain all the data that your camera shot, but they need to be edited and then saved as a different file format in order to be useful to most people. RAW or DNG files may be very large in terms of file size when you take the photo, so keep that in mind. Use RAW or DNG files when you plan to do a lot editing work, and make sure you're using photo editing software that supports these types of files. You'll have an editing learning curve to master to work with RAW or DNG files. For this reason, many drone photographers try to stick with simple JPEG or PNG files.

TIP

If you're just starting to learn about photography, stick with JPEG images. You'll be able to shoot great photos right from the start. After you learn more about editing software, try shooting some RAW or DNG files to experiment with.

Photo resolution

In digital photography, *resolution* refers to the number of pixels in a photo. More pixels mean more detail and color, resulting in higher-quality photos. So, higher resolutions create higher-quality photos.

Resolution is measured using a length-x-width notation. Basic high-definition images are 1920 x 1080 (about two million pixels, or 2 *megapixels*). You can use smaller or even larger resolutions, but consider 1920 x 1080 as the standard (so I wouldn't shoot any lower than this).

Why bring up this topic? You want high-quality photos coming from your drone's camera. You can get the perfect shot, but if the resolution is low, you'll end up with a low-quality image. So you want to make sure you're shooting at least 1920 x 1080, and, honestly, much higher is even better.

Depending on your drone, the default photo resolution may be considerably more. For example, the default photo size for many DJI drones is 12 megapixels (4000 x 3000). That's great, but remember that higher-resolution photos also take up more room on the drone's memory card, so make sure that the card you're using has plenty of storage. (And see the nearby sidebar about moving your photos and video off your drone's memory card.)

TIP

For most drones, you can check and adjust the default resolution on the drone's controller under camera settings. Check your drone's documentation and check out the default resolution setting. You want to make sure you're shooting high quality images!

Shooting Good Angles and Using ND Filters

Good photography is all about composition and angles. After all, as the photographer, you decide what the person looking at your photo will see and the angle at which that person will see it. Angles are especially important with drone photography because you choose the elevation you'll shoot, but you also choose the camera angle you use at that elevation. In other words, you have an additional photography component that a person taking a photo on the ground doesn't have because that person can't fly into the sky. Thankfully, there are just a few principles you need to know and practice in order to shoot great drone photos.

Applying composition

Photography composition consists of choosing what's included in a photo and what isn't. When you take a photo, you consider what the main subject is, how much background you want to include, and whether items are present that distract from the main subject.

Along with the general idea of composition is a principle called the *rule of thirds*. To understand this principle, imagine that a photo is divided from the top and sides into thirds. In many cases, the photo will look better if the primary subject in the photo falls along one of the 1/3 lines instead of the center of the shot. For example, in Figure 17-5, the windmill sits toward the right side of the photo instead of dead center. This composition approach will often give you more pleasing photos, especially when you're shooting a single object.

FIGURE 17-5:
Single subjects often look better when they're not centered in the frame.

However, keep in mind that the rule of thirds is a principle, not really a rule. This principle works well for some subjects, but in many cases, you'll need to fill the frame with the subject. For example, if you're taking a real-estate photo of a home, the rule of thirds probably won't work well. Also, if you need to get close to a subject, such as a tower for an inspection, you'll want to fill the frame to capture the details your client needs.

TIP

The overall composition concept is this: What do you want the person looking at your photo to see — and not see?

Shooting with good angles

If you take a photo of a person looking directly at the camera and then another of that person looking slightly left, you've taken two shots using two different angles. An angle is the degree at which a camera is pointed at a subject. With drone photography, there are a few specific angles that help you get good shots.

Aerial-down angle

An *aerial-down angle* means that you hover the drone over a subject and point the camera directly down at the subject. These drone photos are helpful when you need to show a plot of land or property. For example, a real-estate shoot will often include one or two aerial down angles of the property showing the entire property and its boundary lines, or where the property sits in a neighborhood, as shown in Figure 17-6.

Along with real estate, the aerial-down angle is often used with mapping and agricultural photography. However, keep in mind that the aerial-down angle makes everything look flat. For a general drone photo, it's not the best angle.

FIGURE 17-6:
An aerial-down angle shot.

Aerial eye-level angle

Here's a general photography principle: Don't shoot down at subjects. If you want to take a photo of your dog, don't stand over the dog and shoot down. Instead, kneel down and get on the dog's level. This will give you a much better photo.

This same principle holds true for drone photography. Often, drone photographers hover the drone above a subject and shoot at a downward angle. This approach often distorts the subject and causes you to lose detail. Instead, try to hover the drone at eye level. In Figure 17-7, you're looking eye level at the tall building instead of looking down toward it. This angle gives you a much more pleasing photo that has more depth.

FIGURE 17-7:
An aerial
eye-level
angle shot.

Patryk Kosmider/Adobe Stock Photos

Low angle

A mistake that new drone pilots often make is immediately flying 300 feet in the air and snapping photos. Though it's fun to fly high, often the best drone photos are at lower angles, such as just above the treetops or even just above your head. Think about your subject: Just because you can fly high doesn't mean that doing so will give you the best shot that you're after. This point comes back to the idea of composition, including the question of what you really want another person to see in your photo.

The low-angle shot in Figure 17-8 captures the details of an agricultural field. You wouldn't get the same details if this field were shot from a high elevation. In the end, sometimes the best drone shots are much lower to the ground.

WARNING

Low angles are inherently more dangerous because you can crash into trees, houses, or even people. When you shoot low angles, use the slowest flight mode available and make sure you have collision avoidance features turned on. Watch your surroundings and take your time!

FIGURE 17-8:
A low-angle shot.

Slight downward angle

Although a downward angle often isn't the best, in some cases, a slight downward angle can help you communicate scale and more background detail. This advantage holds especially true for some landscape photos. For example, in Figure 17-9, the cliffs are shot at a slight downward angle to include more background and give the photo a greater sense of scale.

FIGURE 17-9:
A shot taken at a slight downward angle.

TIP

As a general rule, the farther your drone is from an object, the more of a slight downward angle you can use. The closer you are, the less.

Working with ND filters

ND (Neutral Density) filters sound complex, and you've probably at least heard of them in the drone world. In truth, an ND filter is just a sunshade for your camera's lens. That's it. When you take photos or videos with a drone in sunny conditions, you'll battle overexposure because nothing is breaking the direct sunlight. Furthermore, reflective items on the ground may redirect those sunbeams back at your lens. An ND filter helps tone all that sunlight down so that you don't end up with overexposed shots.

ND filters are sold at different strengths and often in sets (think of it as a set of sunglasses with different strengths). Generally, in cloudy conditions an ND8 or ND16 should be enough. For partly sunny conditions, try an ND32. For full sun, start with an ND64 and go up from there. If you're shooting in Pro or Manual mode, the aperture and shutter speed also impact exposure, so keep that in mind. Unfortunately, there's no easy chart telling you what ND filter strength to use; often you have to experiment a bit depending on lighting to find the right one.

You definitely need ND filters if you don't want to end up with a lot of overexposed images. ND filters are sold specifically for your drone because they have to attach to the front of the drone's camera lens, so shop for them using your drone's make and model. (Try the manufacturer's website because they often sell them directly.) After you get a set, start experimenting!

Editing Drone Photos

Most drone photos need a little editing, or *retouching*. Editing just means that you make some adjustments to improve the overall image quality. There are simple photo-editing software products (your computer probably has a basic photo editor built in) and massively complicated software products, such as Adobe Photoshop. Entire books cover the topic of editing photos and using specific software packages, so after you land on the software you'd like to use, a good book to help you learn is a wise investment. For example, Photoshop Elements is a simpler version of Photoshop, and you can pick up the latest edition of *Photoshop Elements For Dummies*, by Barbara Obermeier and Ted Padova (Wiley) for expert help on using the software.

There's a lot to learn about editing, plus it can be quite fun! To get you moving in the right direction, the next sections offer some editing tips to get you started.

Consider cropping

When you crop a photo, you remove portions of the outer areas in order to improve composition. Often, you're just trying to tighten the image a bit or get rid of a distraction. For example, Figure 17-10 shows cropping to remove some of the excess background that doesn't add any value to the subject.

FIGURE 17-10: Cropping can improve composition.

Photo: Curt Simmons

Improving exposure

Most photo editing software has simple tools so that you can improve the exposure of a photo. Did the shot come out a little dark or bit too bright? Exposure tools can often fix these issues for you. Your editing program may provide a one-button Autocorrection fix, with the software making the decision for you. Sometimes this correction is all you need, but sometimes not. You may need to manually adjust the brightness and saturation of a photo to get it right.

WARNING

When you edit, make slight adjustments one step at a time. Too much exposure correction often leaves photos looking artificial, and that's not what you want. Also, don't depend on exposure correction to save your photos. Slight adjustments can help, but always try to get the best exposure when you shoot the photo with your drone.

Improving color

Most editing programs have basic color-correction controls that enable you to adjust the color. Did you shoot a photo with too many yellow tones? You can reduce that tone with a color-correction control. Are the shadows too dark? You can adjust those as well.

As with exposure controls, color-control options can help, but a little goes a long way. Before you begin, spend a moment just looking at the photo. What colors are too dominant? What colors are too bright or dark? After you know what needs to be adjusted, work toward that goal.

TIP

Perhaps you took a nice landscape photo, but it includes a person in the shot admiring the view from the ground. Many editors include a spot removal tool that lets you select the small item you don't want and instantly remove it. Presto!

TIP

If you've used a basic photo editor and are ready to move up a step (but not tackle Photoshop or even Photoshop Elements), consider Adobe Lightroom. It's a nice, intermediate program that is rather intuitive and will do almost everything you may want. You can purchase it with an inexpensive monthly subscription at www.adobe.com.

Chapter **18**

Shooting Expert Drone Video

Your ability to shoot expert drone video will have everything to do with your success as a drone pilot. After all, clients hire drone pilots primarily to gather information. You gather that data with photos, and most often video clips. From agriculture and industrial inspections to mapping and real estate, your video clips will make all the difference. A frequent drone-piloting task is helping clients see what they can't see from the ground.

In this chapter, you get the skinny on shooting expert drone video. You'll consider drone video techniques and delve into file formats and resolution. The chapter also tells you how to shoot high-quality video files using the right drone movements, and I provide some tips for editing your videos.

TIP

As with photography, entire books and courses have been written solely about videography. You might consider picking up a general videography book that goes into more detail. Many of the tips and tricks for videography on the ground also apply to videography shot in the sky. If you're interested in filmmaking in general, check out *Filmmaking For Dummies*, by Bryan Michael Stoller.

Filming Great Drone Video

The best strategies for great drone video actually begin before your drone ever leaves the ground. Determining what you'll shoot (and what you won't) before you fly helps you mentally focus on what you want to shoot. As you think about filming great drone video, keep the tips in the following sections in mind.

Make a plan

My first paid drone gig was shooting a multimillion dollar home for a building contractor. The contractor wanted to use the footage for advertising. I'll admit that I was a nervous wreck. What if I'm terrible? What if I crash into this very expensive home? When I filmed the property, my focus was on flying the drone.

Later, when looking at the footage at home, I was disappointed. I had a bunch of drone movements and good exposure, but something was missing. After my wife watched the video, her comment was, "You're showing me drone movements. Show me the house."

That was a light-bulb moment for me. Instead of thinking about flying, I started thinking like a filmmaker. I wanted to show people the property and make them love it.

So I made a plan and then went back and filmed the house again. This time, the footage was shockingly better.

TIP

The point: To shoot video, you can know all the right moves as you fly your drone, but that takes a means-to-an-end approach. Instead, keep in mind that the footage you're gathering is for people. Show them something. Tell them a story. Always start every drone video shoot with a plan. What do you want people to see? After you answer that question, plan how you will shoot the content and what kinds of drone movements will work best (see the later section, "Shooting with the Best Drone Movements").

Consider the audience

As you make a plan for the video shoot, spend a few moments identifying your audience — that is, who will watch the footage. Perhaps you've been hired to shoot video of a communications tower for inspection purposes. In this case, your audience is internal company personnel who need to see the details of the tower. Slow, steady, detailed video is going to do the trick.

If you're hired to create a video for a beach-house rental, you might use fast, punchy video clips that get potential vacationers excited about the rental.

TIP

Thinking carefully about who is going to watch your video clips will help you make some planning decisions before you ever shoot the first clip.

Modify on the fly

Having a plan is great, but you can't plan for everything. I like to think this way: "Plan carefully, but be flexible." The plans may change, so it's important to adjust along the way. You may be hired by a client who doesn't quite know what they want from the video. The conversation may evolve and change as you fly and shoot. That's fine — just start with your plan but don't become so rigid with it that you can't adjust as needed. Happy clients lead to more work.

Deliver the details

"The devil is in the details," the saying goes, and the details of any process or task will often make or break you, including drone videography. All drone pilots love sweeping video from high in the sky, but in many cases, killer video is lower to the crowd and close to your subjects. That's where the significant details show up.

For example, if you're creating film clips of a house for sale, people viewing that video may enjoy an expansive view of the house and neighborhood from the sky, but what the potential home buyer really needs to see is the exterior details of the house.

As you plan and before you fly, think about the details, both of what you need to show and how to go about doing that. How will you get close to the subject? How will getting close potentially impact safety?

Shoot three times what you need

In many cases, drone footage I thought would be awesome when I shot it just wasn't. When I looked at it on a computer, the footage didn't look as great or interesting as I thought it would.

Here's a truth I've experienced: You're going to shoot footage you don't really like, and sometimes you'll be surprised by clips that turn out much better than you initially thought.

Given that reality, shoot three times as much footage as you actually need. If you need a film clip for a client that's about five minutes long, shoot at least 15 minutes of footage. That way, you'll have plenty of footage to work with and pick through when you edit.

Understanding File Formats, Resolution, and FPS

Before you shoot drone video, you should become acquainted with some technical aspects of video file formats, resolution, and frames per second (FPS). I won't bore you with *all* the technical details, but this section tells you what you need to know to set up these items on your drone controller before you fly.

File formats

A video file format is a kind of file that conforms to certain standards. That file can then be read and displayed by various computer programs. There are two foundational industry-standard video file formats. The good news is that most drones support both of these:

>> **MP4:** MP4 is an industry-standard video file format that virtually every computer and every video editing software package can read and work with. Think of it as the JPEG of video. MP4 videos are very high quality, and you'll be able to use them in any application you want.

>> **MOV:** MOV is an Apple industry-standard video file format. Like MP4, MOV files are very high quality.

On your drone controller, you can choose which video file format you want to use (see your drone's documentation for instructions). The question is which one *should* you use? In terms of quality, there's no difference, so the answer comes down to personal preference.

Here's my advice: If you use a Windows computer for your work and for video editing, use MP4. If you use a Mac for work and editing, use MOV. That's really all there is to it.

Resolution

Video resolution refers to the number of color pixels in each frame of video. The bare-bones minimum resolution to get simple high definition (HD) video is 720p, or 720 *progressive scan* (that's 1280 x 720). But that's the minimum to get HD quality. In reality, you'll want to shoot much higher-quality video than this.

The standard now is *4K video*, which means that the horizontal frame of the video has about 4 thousand pixels (that's 2160p, by the way). Your drone probably supports 4K video and may even support higher resolutions, such as 5K or 6K.

My recommendation is to set your video resolution to 4K, or the highest resolution available on your drone. The more pixels, the more color and definition you'll get, so shoot the highest-quality video you can.

TIP

However, keep in mind that more pixels means higher resolution, which means bigger video files (in megabytes). You need a good memory card on your drone to handle a bunch of 4K video, so don't skimp on the memory card size and quality!

FPS

A video is a bunch of individual frames (think of them as individual photos). Those frames are played in succession, creating the movement you see on the screen. When you shoot a video, a specified number of frames per second (FPS) are recorded. The question is, how many FPS do you need?

TV shows and movies are usually shot at 24fps. Live TV and sports are often shot at 30fps. The general rule for drone video is 30fps. So, ideally, you'll shoot your video clips at 4K, 30fps. This setting will create top-quality video.

However, there's a caveat. Say you shoot a waterfall and, during editing, you want to slow down the action of the waterfall to get that slow-motion look. At 30fps, you may end up with jerky footage. To slow down the action, you actually need more frames per second to get the effect you're after. Think about it: If you shoot 60fps and slow the video speed down by half, you're still at 30fps. That equals good quality and no jerky behavior.

TIP

So why not just shoot at 60fps all the time? Some do, but 60fps can make your video footage a bit darker, and in some cases, the footage can end up looking a bit artificial or overprocessed. The human brain can't interpret a frame rate this fast. Shooting at 60fps gives you silky-smooth footage, but it can also start to look fake. For this reason, most people use 30fps as a default, but then switch to 60fps for footage they may want to manipulate (such as slow motion).

REMEMBER

Don't forget that your drone's video file format, resolution, and FPS are all configured on the remote controller. Get familiar with these settings and where to change them. That way, if you're flying and decide you need to switch from 30fps to 60fps, you can do so quickly while you're still in the air.

Shooting with the Best Drone Movements

Now the fun begins! Because your drone moves through the air, you record video of those movements. You may want to experiment with many drone movements, and deciding on which movements depends entirely on what you're shooting. With that said, certain foundational drone movements tend to produce the best video clips. You should learn to use these movements and put them in your mental tool chest!

Ascend and descend

The ascend and descend movement is just as it sounds: The drone slowly rises up or moves down while filming a subject. This movement is effective for several kinds of shots:

>> **Real estate:** If you're filming a house for sale or under construction, the ascend and descent movements in front of the house are very effective because the movement draws the viewer's eyes to the details as the camera slowly ascends and descends. The motion is much more effective than static video in keeping a viewer's attention on the details of the house. For example, for the scene in Figure 18-1, you could start low to the ground in front of the house and slowly ascend. This would show the viewers the details of the front of the house and then reveal the backyard as well.

>> **Towers and other tall structures:** Ascend and descend movements work great to show the details of any tall structure such as a tower or tall building. This movement can work well for inspections of these tall structures.

>> **Reveal:** The ascend movement can be effective if you want to show the details of a subject but reveal what is behind the subject as well. For example, in an agricultural shoot, you could ascend, showing the details of some crops as you move up, but as you gain height, the footage reveals the size of the entire field. This movement helps communicate not only the details of the crops but also the perspective of the entire field or area.

FIGURE 18-1:
Ascend and
descend
movements are
effective to
showcase the
details
of a house.

TIP

The ascend and descend movements are most effective when you're close to a subject. The movement accentuates details. If you're a long way from a subject, these movements aren't particularly effective.

TIP

Did you shoot an ascent but later wished you had shot a descent? No worries — you can reverse the clip in video editing software.

Orbit

When you use an orbit maneuver, you simply fly in a circle around an object. This feature lets your viewers see the object from all sides. This shot works best with a stationary object that provides details and interest from all sides, such as a building, tower, or house.

When you fly a circle, try to use an elevation that puts the viewer at eye level with the subject, not looking down at the top of it. This elevation provides much more detail and interest.

For example, in Figure 18-2, fly the drone to the elevation and distance so that the portion of the tower is in the view that you want seen. When the distance and elevation are correct, fly and film the circle around the tower.

FIGURE 18-2:
An orbit
maneuver can
show the detail of
a station-
ary subject.

Matthew/Adobe Stock Photos

Many drones have preconfigured flight paths, such as circle patterns and other basic movements. You can have the drone fly this path automatically with just a click of a button. See your drone's documentation for details.

TIP

When you edit orbital footage, you can use a normal speed for half the circle and then speed up the footage for the second half, which can create an interesting and dynamic video clip.

TIP

Semicircle

Like the orbit movement, the semicircle movement is similar, but you show only half the circle. This movement works well for construction sites, real estate, and other buildings where showing the front is most important.

This movement is effective because as you make the arc of the semicircle, the drone movement creates depth that you wouldn't get if you just fly in a straight line. This depth can be visually pleasing and reveal more of the property or area surrounding your subject. In Figure 18-3, a semicircle flight path would do a good job of showcasing the front of this home.

FIGURE 18-3:
A semicircle path
adds depth to
this image.

Stephen VanHorn/Adobe Stock Photos

TIP

Here's a quick tip: Your drone may not have an automated flight path for a semi-circle. If that's the case, just fly a complete circle and then edit the footage afterward to show only the semicircle portion you want.

Bird's-eye rotation

With a bird's-eye rotation, you hover your drone over a subject or area and slowly rotate the drone in order to add interest to your video. This movement works well when you want to showcase an entire neighborhood, a farm, a shipyard (shown in Figure 18-4), a construction site, or the boundaries of a piece of land.

When you film this maneuver, go slowly. Although this maneuver adds interest, a little goes a long way, and moving too quickly can be both annoying and dizzying for your viewers. To film this move, simply put your drone in the position and at the elevation you want, point the camera down, and slowly make a 360-degree rotation as you film.

Fly in/pull back

This simple drone move can be very effective when you want to showcase a single subject, such as a house, a boat, or a tower but also want to reveal some of the surrounding area.

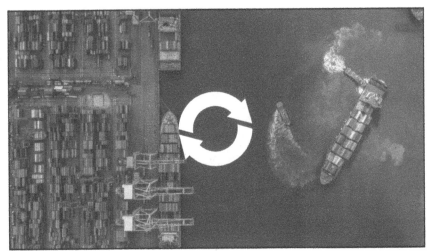

FIGURE 18-4:
Use the bird's-eye rotation movement to add interest to a big subject.

MAGNIFIER/Adobe Stock Photos

The fly in/pull back feature (sometimes called "dronie") are two separate movements with the same goal: Start close to a subject and then pull back away from it, or start far from a subject and fly toward it. The odds are good that your drone has an automated movement for a fly in/pull back or something similar. Using the built-in maneuver, the drone will likely reduce elevation on the fly in and gain elevation on the pull back, as shown in Figure 18-5. The elevation changes also add a nice feeling of vertical movement to the video.

REMEMBER

Did you shoot a fly in that you wish was a pull back, or vice versa? Don't worry — you can reverse the clip in a video editor.

FIGURE 18-5:
Fly in/pull back provides details of a subject and the area around it.

onuma Inthapong/Getty Images

Sweep

In the sweep maneuver, you fly in a straight line at the same elevation across the length of a subject. This move works well to show the details of a building or house that is horizontally big. With this move, you simply fly close to the subject and move from one end to the other in order to reveal the subject. For example, a sweep of the house in Figure 18-6 would reveal the details of the house as the drone flies a straight line in front of it.

FIGURE 18-6: Use the sweep move to reveal the details of a large, horizontal subject.

TIP

This movement works best when you fly "low and slow." After all, you're showing the viewer the details, so don't rush it.

Track

The odds are good that your drone has a tracking feature. Such a feature just means that your drone can automatically follow a moving object, such as a car, a boat, someone riding a bicycle, and so forth. Tracking a subject helps communicate movement and adds interest to a video. For the right subject, tracking is very effective.

Depending on your drone's capabilities, you can probably follow and track a subject moving away from or toward you (the drone flies backward), or you can fly parallel to a moving subject. Check your drone's documentation for information about setting up and using these automated features.

WARNING

If you fly parallel, your drone will fly sideways. Many drones have front and rear collision-avoidance sensors but not side sensors, so if you fly parallel, you may not have any crash protection. Keep that possibility in mind and use caution.

When you track a subject, think carefully about distance and elevation. These shots are often more effective if you're closer to the subject and therefore closer to the details. Also, flying at a slight elevation above your subject is often effective, as shown in Figure 18-7.

FIGURE 18-7:
Keep closer and fly at a slightly higher elevation.

pixdeluxe/Getty Images

TIP

Time is often the enemy. When you shoot video, you may have limited time. That's why having a plan in place is so important; you don't want to forget any important shots. Also, drone batteries don't exactly last a long time, so you have to keep things moving. Practice flying these maneuvers. The more you practice, the better and faster you'll be!

Getting Started with Video Editing

To produce effective video content for your clients, you need some video editing skills. Happily, you don't have to know everything there is to know about video editing. Most people can produce good video with just a few editing skills, so don't think you have to be an expert. Learn what you need to know to get started, and then learn as you go. That's the best way.

The first question people ask is what video editing software to use. Most video editors today can handle 4K video, so you have several options. Here are a few:

» **Built-in software:** Both Windows and Mac computers have basic video editors built into them (or you can download them for free if not). These editors are a good place to start, but they don't do a lot, and you may find their lack of options frustrating.

» **DaVinci Resolve:** DaVinci Resolve is a great editor that works on both Windows and Mac. You can find tons of features in the free version, so it's a great way to step up your editing game without spending money up front. Find out more at https://www.blackmagicdesign.com/products/davinciresolve.

» **Adobe Premiere Pro and Apple Final Cut Pro:** For full-service, professional-level editing, Adobe Premiere (https://www.adobe.com/products/premiere) and Apple Final Cut Pro (https://www.apple.com/final-cut-pro) are two of the most popular programs. There's a lot to learn with these, but again, you can take it one step at a time. Also, you can download trial versions to get started.

TIP

After you land on an editing program you want to use, you have many books and training programs to consider. These options are very helpful because these complex programs are not always intuitive.

Before you get into the weeds of a software program, it's important to understand the *workflow*, which refers to what you do with your video clips and in what order. For the details on using the program you choose, you'll need to check out other sources. For the big picture in editing drone video, read on.

Importing video clips

The first and most obvious step is to get the drone video clips off the drone, onto the computer, and then imported into the program. Most video editing programs have a *storyboard* or some other holding area where you import your clips. After you have everything you need imported into the program you need, you're ready to get started.

Editing each clip on the timeline

When you have your clips in the program, start by adding one clip at a time to the *timeline*. Your program may then have to *render*, or process, it before you can do anything with it, and this process can take some time. When the clip is ready, do this:

» **Trim:** Trimming is the process of deciding what you do and don't want to keep, so carefully watch the clip and start trimming superfluous or low-quality areas of it.

>> **Split:** Split clips into different parts as needed. For example, you may have a longer video clip with several pieces or drone movements. You can split that clip apart so that you end up with two or more clips. This way, you can work with each clip individually and reorganize clips on the timeline in any order you want. For example, in Figure 18-8, I split one clip into two as a way to break apart two different scenes recorded in one video clip.

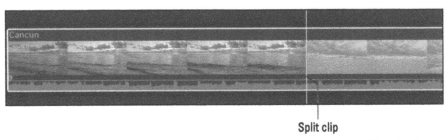

Arrange clips and apply transitions

After you have all the clips you want to use, as described in the preceding sections, you can organize those clips on the timeline. In most programs, you can simply drag clips around until you get them in the right order. Then you can apply transitions to the clips, such as crossfades, if you want to use them. (A *crossfade* blends one clip into the next.) Now is a good time to look at the timeline and see how long the video is. Do you have a five-minute maximum for this video, but seven minutes are on the timeline? If so, it's time to trim some more.

TIP

Get everything correct and in order before you move forward. Doing so now will save you time and aggravation later. As you work, be critical. Make sure that you like every clip and don't need to trim anything more. Sometimes it's helpful to take a mental break at this point. Give yourself a bit of time, and then come back and look at all the clips again.

TIP

As you work with video clips, keep in mind that most programs will let you change the speed of a clip (faster or slower), and you can even reverse the clip so that the action is moving in the opposite direction. These tactics can be very helpful when editing drone footage. Often, using the tools you need requires simply clicking a button in the program, so see the program's documentation for details.

Edit for color and exposure

Most editors have some color correction and exposure correction controls that are automated. Sometimes these help a lot, and sometimes not so much. You may end up having to make manual adjustments, depending on the tools that are available in the program you are using. Two tips:

>> **A little goes a long way.** Too much color or exposure correction often leads to fake-looking video clips. Make small adjustments, one at a time. If something doesn't look right, you can then back out of your last adjustment. If you make a bunch of edits simultaneously, it's hard to backtrack from them.

>> **Watch out for clip inconsistency.** You want the video clips to form one movie. To accomplish that, your color and exposure correction will need to be similar for each clip so that one clip isn't too bright and one too dark while another is too brown and another too green. As you look at the details of a clip, keep in mind that the clip is a part of the whole.

COLOR GRADING AND LOOKUP TABLES

If you read much about editing photos and videos, you'll probably stumble across the terms *color grading* and Lookup Tables (LUTs). In short, color grading is the manual process of applying color profiles to an image or video. Although color correction helps you get the best real-life color, color grading can help you get a visual tone that's pleasing and rich. However, color grading is complicated and involves a learning curve (entire courses exist just for color grading).

As a shortcut, you can use a LUT. A *LUT* is a file that provides a set of color tones you can apply to a photo or video. In more advanced programs, like Adobe Premiere Pro or Apple Final Cut Pro, you open your video, open a LUT file you want, and then apply the LUT and adjust the effect percentage. However, finding the right LUT for a particular video can be like searching for a needle in a haystack. After all, you want the right effect and the right color tone. Yet, if you have a collection of videos all shot in the same location and you find a LUT that makes them look great, you've improved the collection. If you're interested in experimenting with LUTs, many websites sell them. Get started at Thelutbay.com to find out more. Thelutbay.com even offers LUTs specifically designed for drone photos and videos.

Add music, voiceovers, and other effects

After you've assembled your video, you're ready to add music or voiceovers. Typically, you just drag these kinds of files to the timeline, adjust the location, and adjust the volume. You can also add titles, credits, and other text effects throughout if needed. Just check out your software documentation for details.

REMEMBER

Keep in mind that almost all music is copyrighted, and you can't use it for commercial purposes, even on social media. You'll need to use music that is licensed for commercial use. Your editing program may have some built-in, free tracks you can use. For much more diversity, most filmmakers subscribe to a music service so that they can download and use royalty-free music in their projects. The following are a few services to consider:

>> **Epidemic Sound:** A great resource for royalty-free music as well as sound effects. Visit https://www.epidemicsound.com/ for subscription details and pricing.

>> **Soundstripe:** Also provides plenty of royalty-free music and sound effects for an unlimited annual fee. Go to https://www.soundstripe.com/ to find out about the subscription options.

>> **PremiumBeat:** This Shutterstock company offers *extensive* royalty-free music options through individual purchase or annual subscription. Get the details at https://www.premiumbeat.com/.

Finalize, export . . . and then review!

All right, you've made it to the end of editing your video! Now play your entire movie in the timeline, noting transitions, music and voiceover volumes, and any text or other effects you've added. Make sure everything lines up in the places you want on the timeline. When you're done, just export the movie.

Now comes the hardest part. It's time for another person or two to watch your video. Sometimes this feels like sending your newborn baby out into the world, but you need other people to review your work. Ask them to be critical and point out everything that seems "off." Make a list of those items; then go back to the timeline and make adjustment. Next, re-export and repeat this entire process. It may take several export passes before you're really satisfied with your video. That's okay! After all, this, too, is a part of the editing process.

Chapter **19**

Getting Drone Piloting Jobs

or many people, drone piloting begins as a hobby. It's fun, and you have a good time flying your drone. But there may come a point when you ask yourself, "Can I make money with my drone?"

The answer is yes! Drones are used in all kinds of industries for different purposes, and there is always a need for skilled, FAA-certified drone pilots. You just have to know where to look and what piloting jobs appeal to you.

Think of this chapter as a crash course in drone piloting jobs. You explore some industries that commonly use drones, and I offer tips on how to find out whether pilots are in demand in your area. Also in this chapter is guidance on marketing your business, figuring out what to charge for your services, getting paid, and dealing with insurance and taxes.

Do you want to launch out on your own and open a drone piloting business? That's what I did, and I give the skinny on how to go about that as well!

Exploring Drone Piloting in Business

Drones have made business tasks easier and, in many cases, safer. For example, in the past, if a Realtor needed an elevated shot of a house, they needed something to stand on, and they just couldn't get the great shots you can take now. Drone piloting has likewise improved many other industries by providing information from the air.

The next sections offer a look at some of the major categories of businesses and industries that make use of drones and specific drone piloting skill sets

Real estate and mapping

The industries that tend to employ the most drone pilots are real estate and related marketing along with mapping. In today's market, potential customers expect aerial photos of a home, land, and businesses that are for sale. In fact, it's standard to include shots that show the property and boundary lines from the air. In Figure 19-1, a drone was used to mark the boundary lines of land being sold for future home construction. Without this photo, the potential customer can't visualize the scope of the property.

FIGURE 19-1:
An aerial view of real estate.

Bigc Studio/Adobe Stock Photos

In some cases, real-estate agents get their Part 107 license and shoot the photos and videos themselves. However, in most cases, real-estate agents hire a freelance drone pilot. A freelancer is someone who owns their own drone piloting business and is hired to shoot photos and videos on a project-by-project basis.

Real-estate photography and videography may be something you should consider. If the real-estate market is booming and turnover is quick in your area, you can stay rather busy after you get your name established. However, in many cases, real-estate agents prefer drone pilots who are also photographers who can shoot the interior of the house as well, rather than hiring two different people. So to excel in this business, you may need a professional DSLR setup and the skills to shoot inside the home or property. This endeavor consists of an entirely different skill set, of course. If you're interested, YouTube.com is a good place to start because you can find numerous videos that will teach you the basics of shooting interior photos.

TIP

The good news is that you don't have to spend a fortune on a drone to get started in real estate. Most drones with a decent camera can shoot sufficiently high-quality photos and videos for real-estate marketing purposes. Drones with numerous cameras (such as the DJI Mavic 3 Pro) are great, but for real estate photography, you don't need something that advanced.

However, you can learn and offer more advanced photographic work for real estate, such as mapping. At a basic level, drone mapping creates a map of a large area by using many individual photos. You then put together those photos, which are based on GPS coordinates, to form a single image. You use mapping software, such as Pix4D, Drone Deploy, or DJI Terra, to assemble the images.

In more complex mapping projects, you can create 3D models of property, structures, and landscapes using a process called *photogrammetry*. These 3D models may be used in commercial real-estate applications or surveying. Commercial-level drones use LiDAR (Light Detection and Ranging), a laser scanning feature that creates 3D models of everything from landscapes, structures, and even the interior of large buildings. For example, Figure 19-2 shows you a LiDAR-scanned model of an urban area, giving you a 3D look at the land's surface features, or topography

Of course, photogrammetry and mapping use a specific skill set that you'll need to learn if you don't already have it, and you'll need drones capable of LiDAR and software to produce the models. Also, you may need special permits or licenses. Still, depending on the market where you live, this skill set and equipment may be quite lucrative.

Agriculture

The agricultural sector is one of the fastest-growing drone piloting industries. Aside from using a standard drone to get an aerial look at fields of crops, numerous drones have been developed that can do much more. For example, today's agricultural drones can do the following:

>> **Seeding:** Drones can be used to map and seed fields with great precision and much more quickly than traditional methods.

>> **Spraying:** Drones can be used to spray fields for pests and diseases and even to distribute fertilizer.

>> **Multispectral imaging:** Multispectral-enabled drones can take a variety of photo types and help farmers see areas of fields that have growth problems so that those areas can be targeted for treatment. The camera in this type of drone uses a layer-by-layer approach and can build images showing specific light wavelengths, enabling farmers to see areas of a field that might not be visible with a standard photo.

>> **Irrigation:** Some agricultural drones can detect ground moisture and identify areas that may need irrigation. This feature enables farmers to see areas that need attention before crops are damaged or lost because of a lack of water.

>> **Livestock tracking:** In large-scale livestock operations, drones are used to track livestock numbers and movement patterns, and even to quickly locate missing livestock.

>> **Weather monitoring:** Some drones can track air temperature, humidity, wind speed, and wind patterns. These tools can be used to predict weather patterns and possible weather-related problems.

These are just some of the potential agricultural drone applications and uses. An example in this industry is the Hylio AG-230 drone, as shown in Figure 19-3. The AG-230 is a drone designed for crop spraying and can be used as a single drone or in a swarm of drones where the drones coordinate and work together to cover a large area. A single AG-230 can spray up to 50 acres in an hour (based on flow rate), has obstacle avoidance and first-person view (FPV), which involves wearing goggles with a built-in screen, and uses radar technology to maintain a precise distance over obstacles and trees.

FIGURE 19-3:
The Hylio AG-230.

Photo courtesy of Hylio

Law enforcement

Law-enforcement drones have had a big impact on public safety because they can be used in many different ways. For example, drones specific to law enforcement can be used to view and reconstruct traffic accidents and access areas that are difficult for humans or are unsafe, and are used by SWAT teams or in hostage negotiations.

Law-enforcement drones are often outfitted with infrared camera technology so that they can see in both light and dark situations. One of the most important keys for law enforcement is that the drone can provide real-time information so that law enforcement officers can make the best tactical decisions.

An example of a law-enforcement drone is the Brinc Drones Lemur 2, shown in Figure 19-4. This drone is designed to keep people safe with tools to aid in HazMat (hazardous materials) response, search and rescue, hostage negotiation, and other applications. The drone has 4K day and night vision and thermal sensors. It can enter structures and create real-time, 3D interior maps (using LiDAR), and 2-way communications so that law enforcement officers can talk to people through the drone. It even has a glass-break attachment, enabling it to break through a window and enter a structure. Should a person knock the drone to the ground, it can right itself and begin flying again. The Lemur 2 has a 360-degree position hold and can hold this position without GPS or light. Along with these features, the drone can even carry small items, push through interior doors of a building, and provide real-time video streaming to officers.

As you can imagine, features like this provide detailed information and help law enforcement officers stay safe in difficult situations.

FIGURE 19-4:
The Brinc
Drones Lemur 2.

Search and rescue

Along with law enforcement, some drones are specifically made for search and rescue operations. These drones are typically outfitted with night vision and thermal imaging and can search remote areas, complex or hard-to-access topography, or the interior of buildings. They are also used to find people or disabled boats at sea.

The DJI Matrice 350 RTK and the 3DR Solo are leading examples in this category.

One of the most important aspects of search and rescue is thermal imaging, which can help teams locate people who are lost or may be trapped in disaster areas. In Figure 19-5, a thermal-imaging drone is inspecting the site of a fire, looking for people and potential hotspots. Without the thermal imaging, the smoke and obstructions would make this search impossible.

FIGURE 19-5: Search and rescue using thermal imaging.

Construction and inspections

The construction and inspection industries are big fields for drone pilots with many different applications. In construction, drones are used to map and coordinate large-scale construction projects, record progress based on timelines, and provide security and potential hazard information. Along with site mapping and surveying, drones are essentially used to provide information to help make construction decisions as well as avoid potential pitfalls or dangers.

For example, in Figure 19-6, the drone is photographing damage to a pipe at a facility. The drone can get close to the damage details so that inspectors can make repair plans.

Speaking of inspectors, they use drones constantly to inspect everything from towers, buildings, and building interiors to hard-to-reach points, including one type of area you might not expect: sewers and tunnels. Though many drones can be used to inspect items and photograph them, there are specialized drones designed to enter sewers and tunnels that may not be safe for humans.

For example, the AeroLion BlackLion 078, shown in Figure 19-7, is specifically designed to inspect sewers and drainage tunnels. This drone can withstand humid environments, work in small spaces, detect potential tunnel defects and leaks, and float on top of water. The drone has autonomous flight and positioning and can automatically take photos based on distance. It can even generate an inspection report.

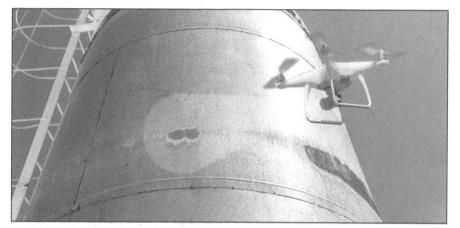

FIGURE 19-6:
A drone used to
find and
photograph
damage.

Parilov/Adobe Stock Photos

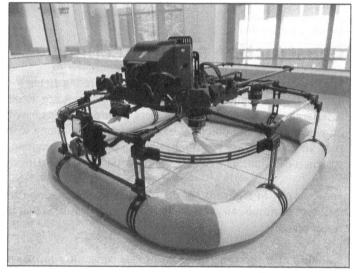

FIGURE 19-7:
The AeroLion
BlackLion 078.

Photo courtesy of AeroLion

Shipping and logistics

A developing field in the drone industry is shipping and logistics. Drones can be used to carry and deliver packages, which saves an enormous amount of time. Drones can also be used in inventory management and order picking inside warehouses. Although much of this industry is still developing at the time of this writing, companies such as Amazon and Wal-Mart are working to use and deploy drone delivery systems.

Flytrex, DroneUp, and Zipline are just some of the delivery companies developing and providing drone delivery services. With these services, even groceries and medications can be delivered with a drone.

Delivery drones are outfitted with different kinds of mechanisms to hold and deploy packages, as shown in Figure 19-8. Of course, weight is an issue, so don't expect your $400 grocery bill to be a drone delivery option. Still, many items, such as time-sensitive medications, can potentially be delivered very quickly. As drone delivery systems develop, I believe we'll see more and more drones used to deliver all kinds of items in the near future.

FIGURE 19-8:
A drone outfitted with package transport and delivery mechanism.

chiew/Adobe Stock Photos

Film and photography

Naturally, the film and photography industries use drones for everything from aerial photography to making movies. The key for these industries is, of course, the drone camera, so all kinds of drones are used in these fields. Pro-level photography drones, such as the DJI Mavic 3 Pro and the Autel Evo Max 4T, are just two examples. As the cameras and prices climb into the sky (so to speak), cinematic drones such as the DJI Inspire (one of the most popular) are put to work creating the content we all enjoy.

Security

Finally, many drones are used for security. From simple surveillance and subject tracking to video mapping and all related applications, security drones are used by private companies all the way up to border protection. They're even used in wildlife conservation efforts to deter poaching.

The security sector uses numerous drones depending on the kind of security needed. For more advanced features, drones designed specifically for security include the Kespry 2S, Microdrones MD4-3000, and Easy Aerial SAMS. Each of these provides different features for different kinds of security operations.

Finding Your Drone Piloting Career

As you get ready to move toward a drone piloting career, it's not always easy to know how to begin. In this section, I try to get you moving in the right direction by looking at your current job, area of interest, and current skills.

REMEMBER

Remember, you must have your Part 107 license to do anything commercial — whether you are paid for the job or not. So moving forward, this discussion assumes that you already have your Part 107 or are in the process of preparing to take the test.

Looking at your current job

One of the best ways to get started in drone piloting is to look for any opportunities in your current career. Maybe you're currently working in a job that doesn't offer any opportunities for drone work, but think carefully. You might be surprised!

If you work in construction or a related industry, there may be a need for photos and videos of projects in process. Providing such photos and videos is a great way to get your foot in the door and start building some drone piloting experience. Even small construction firms can benefit from aerial photos and videos for advertising purposes and even just to use on social media. Don't let the size of your company dissuade you; most companies have some kind of social media presence, and drone footage of projects can only help.

The education and nonprofit industries often have different kinds of events that make great drone photos and videos. If you're a teacher at a school or a leader in a nonprofit organization, you may have more of an opportunity than you think to take aerial photos of events.

REMEMBER

The point is not to write off possible opportunities at your current job. Even small businesses often need photos and videos for promotional purposes to generate the attention of customers.

The big ask

Here's the thing: If you work for any kind of business, there's a good chance your manager may not realize that drone photos and videos can be an asset. This reality simply may not have occurred to your manager.

So, ask. You can simply say, "I have my drone piloting license from the FAA. Are there any opportunities for me to take photos and videos to use in advertising or our social media? Because I'm certified, anything I shoot is legal to use."

TIP

In bigger companies, you may be able to negotiate some new job responsibilities and perhaps even a raise, but your manager has to know about your skills and certification. Just bring it up! Managers always appreciate employees who want to be involved and help a company do more, and your drone may be able to do that.

REMEMBER

At this point, the question of money always comes up. Here's what I recommend: You want your foot in the door. Achieving that goal is more important than making money at the beginning. Do some shoots for free. Show the people with whom you work what you can do first. Then you can always ask for pay or a stipend for future projects. Even free shoots help you build a portfolio of work, which can only help you in the future. You're trying to get yourself established as a drone pilot. Work on that goal first and money will tend to follow.

A second big ask

Sometimes a business you work for is willing to help if they see value in having a drone pilot on board. They may be willing to pay for your Part 107 certification or give you a stipend to help you upgrade your drone equipment. Just keep it professional, friendly, and casual and say, "I can do some photo and video shoots at no charge, but can you help me with certification costs or equipment costs?" Even a little extra money coming in can help you get that new drone you want! It doesn't hurt to ask, and often a business will help an employee who provides a valuable service.

Exploring your area of interest

As you get started in drone piloting, you're likely to shoot photos and videos. But as you begin doing projects, it's important to think about your areas of interest. Is that what you want to do, or are you interested in more complex drone work, such as photogrammetry? (I explain photogrammetry in the "Real estate and mapping" section, earlier in this chapter.) Or you may be interested in complex inspections, or security, or search and rescue.

TIP

Think about what you would actually like to do, and then do some online research. Consider whether you need to take additional classes or get additional certifications. Work on building your business skills and taking small photo and video jobs, but always keep an eye on your goal and work to get there. You can't do everything there is to do with your drone, so it's worthwhile to determine what you truly want to do.

Developing skills as you work

New drone pilots often think, "I've got to have perfect skills before I accept any drone piloting job." That's not true!

Sure, you need to know how to fly and manage your drone in a variety of situations. Be sure to practice the maneuvers you find in Chapter 3. Use obstacle avoidance and don't use sport or speed mode (typically, the drone sensors don't work in this mode). Slow down and think. Understand and practice taking and editing good photos and videos (see Chapters 17 and 18).

TIP

If you're doing these things, then go ahead and work on getting a few drone gigs that are simple and straightforward. Hands-on practice is one of the best ways to continue developing your skills as you work for other people. Good drone pilots fly a lot and practice flying in different situations. The more you fly, the more skills you'll gain. Small piloting jobs that aren't overly complicated are also a great way to grow your skills. So get out there!

Setting Up a Drone Piloting Freelance Business

Although you may be able to pilot your drone for an existing business or organization where you work, many drone pilots eventually launch their own small business. A small drone piloting business doesn't have to be complicated, but many people often miss some critical steps to success.

The first tip I offer is this: Don't quit your day job! A freelance business is great, but making a profit may take some time. Any business takes time for development and consistent earnings. So again, don't quit your day job; instead, work on your freelance business on the side and grow it over time.

Defining your business

My drone business specifically focuses on taking drone photos and videos for individuals or other businesses. I tend to shoot for a nonprofit, construction businesses, and individuals with property. That's it. I don't do anything else and don't want to.

Especially at the start of your business, you may be tempted to take any job that comes your way, even if it's outside your interests (and, honestly, current level of skill). Taking any job you get offered can lead to a lot of frustration over time. I encourage you to think about what you really want to do and focus your business on that goal. People who focus on a goal tend to reach that goal. If you're pulled in too many different directions and take on jobs you really don't want, your business goal can easily drift.

REMEMBER

With that said, it's important to note that sometimes piloting job opportunities may come your way that are outside of your main goal, but also interesting to you. Sometimes you may end up doing a piloting job you weren't pursuing but actually enjoyed. This experience can lead to more work in an area you didn't expect. So be open to new opportunities, but also remain focused. You're looking for a balance between the two. Remember, defined business goals are attainable goals.

Setting up your business

A drone piloting business typically has one person — you! So it's not hard to get a business started because you're a single freelancer. It's not like you need a bunch of employees to get your business off the ground.

With that thought in mind, some drone pilots simply get their license and work on getting a few freelance jobs. You make a little money on the side, have a good time, and for many people, that's enough.

TIP

However, to establish more than a casual business, you probably want some kind of business structure. This structure helps you operate as an actual business and helps you make sure your taxes are paid (more about that fun issue coming up in this chapter).

For most drone piloting businesses, you'll likely form one of two kinds:

>> **Sole proprietorship:** This is the simplest type of business, and you may not even need to formally organize it with your state. You may need to file your business with the county you live in, especially if you use an assumed business name ("doing business as," or DBA). A sole proprietorship is sort of like saying, "I have a business in this county," but that's typically about it.

>> **Limited Liability Corporation (LLC):** Many small drone businesses form an LLC, even though only one person is involved (you). An LLC is a way to shield yourself and your family from any financial or civil liabilities. In other words, if you crash your drone into the building and damage it, that company can sue your LLC, but not you and your family personally. In fact, many larger companies won't hire you for drone piloting jobs if you're not an LLC, so it's something to consider. An LLC is formed in the state where your business resides and filed with the Secretary of State.

The best place to begin setting up your business is with an attorney. Typically for a reasonable fee, an attorney can look at your business and help you make the best decision about what type to form. The attorney may also be able to help you form an LLC with your state. So it's good to get some legal advice for your context. Also, you may be able to pay a *registered agent*, who is someone who takes care of the registration process to help you set up your LLC. My small business is an LLC, and I paid a registered agent to get everything set up with that state and provide a few other business services. Honestly, I could have done it myself, but it was worth the money to just pay someone else to take care of it. Just do an internet search for "LLC registered agent *state*." Because LLCs are formed at the state level, you want a registered agent for your state.

Although registered agents can help, if you're new to owning a business, I recommend talking with an attorney first if you have the funds. The attorney can help you get things organized and make sure you understand the business structure you're entering into.

Making first contact

No, we're not talking about aliens here; we're talking about customers! When I first started my LLC, I contacted people I already knew: real-estate agents, construction contractors, and a few individual businesses. I offered them greatly discounted rates to get the ball rolling. I was then able to use some of the photos from those projects in local advertising.

The point is, when you first start, you need to get some momentum going. Think of your business as a train sitting still on the tracks. It takes a lot to get things moving, but after you're moving, it's easier to keep moving.

Your best starting place is with people you already know. Think of anyone who may need drone work and offer them a big discount just to get your business started.

Establishing an internet presence

In today's business climate, you need a web page and social media presence. When people look for a business, the internet is the first place they go. Though the details of establishing an internet presence and marketing your business is worthy of an entire book in itself, following are some tips to get you moving in the right direction.

Establish a domain

An internet website requires a domain name, and you'll need to choose that domain name and register it. Many companies will do this for you. I've used GoDaddy.com and can recommend it. On that site, you can search for a possible domain name you'd like to use.

Ideally, the domain name should be your business name, but that can be complicated because the domain name has to be unique. For example, my business name is Sky Life Photography, LLC, and the domain name is http://www.skylife photography.com. Sometimes small businesses will find the domain name they want before establishing a business name because getting the domain name can be difficult. Each domain name must be unique, which means that someone else may already own it. In that case, you have to find a variation of the domain name or choose something altogether different.

REMEMBER

Whatever you choose, keep it as simple as possible. You want to be able to tell someone a website address that's easy to enter in a browser. Avoid long, confusing addresses or complicated abbreviations. Simplicity wins here, so think carefully!

Build a web page

You can pay someone to create a web page for your business, but with the right platform, you can probably do it yourself. Many companies, such as Squarespace. com and GoDaddy.com, provide website services with templates so that you can just insert your content without knowing much at all about website creation. With good templates, you can build a good-looking website in less than half an hour.

REMEMBER

With that last thought in mind, here are a few quick tips for your drone piloting website:

>> **Keep it simple.** The more complicated your website is, the less people will look at it. Keep it as simple as possible with minimal reading required.

>> **Organize the pages.** Have several pages on the site that provide direct and easy-to-access information. For example, have a page about your services and one about pricing. Make it simple, organized, and clear.

>> **Use images.** Use your best drone images and short videos to showcase your work. Don't include everything you've ever shot; instead, choose the best examples that highlight what you can do.

>> **Keep it modern.** You want the web page to look clean and modern, not one that looks 20 years old. You sell technology services, and your website needs to look modern. This is why templates matter. They're already set up for a modern look.

>> **Use contact forms.** Make it easy for people to contact you. Include a phone number and email address but use forms on the web page so that users can quickly fill out a contact form directly. That's what people expect of modern websites.

Set up a social media presence

Facebook and Instagram can be great friends for your business. You can probably tie the two accounts together so that posting to one also posts to the other. Your social media is a place to find customers and showcase your work.

When you set up your social media, make sure your contact information is complete and easily accessible. Keep a close check on the Messenger app because people may contact you that way rather than via phone or email.

Also, update the site with new information, photos, and videos regularly. Depending on your business and its location, you may need to post something a few times a week, or perhaps only once a month. You want to be engaged as much as you can. But don't make it so complicated that social media becomes a job in and of itself. I live in a rural area and don't have much business competition. I use social media, but just enough to stay engaged without letting social media take over my life.

TIP

Check out *Social Media Marketing All-in-One For Dummies*, by Michelle Krasniak and others (Wiley) for comprehensive guidance. Other *For Dummies* books about social media marketing, starting a small business, and more are also available at www.dummies.com or on Amazon.com. You can also find all kinds of instructions, articles, and videos on the internet about establishing a social media presence. Just search for that content and get started.

Advertising your business

Hear this: Word of mouth is your best advertising. It's free and the most effective kind of advertising. The more people know about your business, the more they tell others as the topic comes up. Your customers are your best source of new customers.

With that said, you may also need to advertise your business to attract new customers. You have a few different ways to advertise:

>> **Newspapers:** Yes, in some areas, newspapers are still a great way to advertise.

>> **Billboards:** You may be able to find an affordable billboard spot or share a spot on a digital billboard.

>> **Sponsorships:** You may be able to sponsor a local sports team or organization. This can help get your business in front of many people, and your community will thank you for sponsoring a team.

>> **TV:** You may be able to run a local TV ad that's helpful and affordable.

>> **Social media:** Social media ads can be a great way to advertise your business to new people because social media is all about connecting people! Facebook, Instagram, X, LinkedIn, YouTube, and Pinterest are some of the most popular.

What I want you to remember is this: What works varies. It depends on where you live and your community. What works well in one location may not work well in another. In my area, social media ads have been the best way to get my business name out there. Look at what kind of advertising other local businesses use and follow suit. You may have to experiment a bit with some types of ads and different media to discover what works best for you.

REMEMBER

If you use social media, don't forget that social media ads can be targeted to your area. I can't tell you the times I've seen ads for local businesses in my social media feed that are 500 miles away. That's a huge waste of money, so make sure you carefully target the ads to your service area.

Getting paid

One of the biggest questions drone pilots ask on internet forums is, "How much should I charge?" There's no good answer because it depends. It depends on what services you provide and where you live. The charge drone pilots can ask for a particular job varies wildly depending on the current market in your location. Read on for more details about charging for your work and getting paid.

How to charge

TIP

To determine charge: Find out what other drone pilots are charging in your area and get in the middle of the pricing. Don't be the lowest, but don't be the highest, either. Getting in the middle of the pricing range is your best option and will lead to more customers.

Also, sometimes you can't do a job for the pay, so be willing to say no when you have to. For example, one time I was asked to do a real-estate shoot two hours away from home. I just can't charge enough to overcome the gas expense and my personal drive time for a job that far away, so I had to turn it down. Another time, I was asked to shoot a graduation party, but the customer wanted DSLR images rather than drone images. Though I shoot with my DSLR, I don't do group events of people, so I turned it down. I always hate to tell customers no, but I also can't lose money, and because I want to stay focused on my business goal, sometimes no is the best answer.

When to get paid

REMEMBER

Unless you have an established, trusted customer who pays after invoicing, you should make it clear that payment is expected at the time your services are rendered. Do not provide photos and videos to a person who hasn't paid you. Get the payment first, and then provide the content you created for the customer. You might be surprised at how many people won't pay you after the fact, so get paid up front.

ALWAYS FOLLOW-UP

Never ignore emails or messages from a customer. If you receive a request that's outside of what you provide, answer with a polite declination and refer the customer to someone else if you can. You never know when that customer will come back to you with a job that aligns with your business plan. If you don't respond, you'll never get that customer in the future. It's shocking to me how many businesses just don't respond to a potential customer. That's terrible business practice and will hurt you in numerous ways. In fact, that person will probably tell other people that you never responded! So always respond to every email or message with a professional, courteous response. Your drone piloting business will thank you in the long run!

Juggling income and expenses

One of the hardest things about a small business — and one of the greatest errors small business owners make — is juggling income and expenses. Here's the overall thought: Log everything.

You'll need to log all your income, of course, but often small expenses are overlooked, and you may not be making as much money as you think. Also, as a business, you can deduct the expenses from your income for tax purposes, so you need to keep meticulous records and be able to document every income and expense you have.

With that said, here are some tips to get you moving in the right direction for managing your income and expenses:

» **Consider software.** You may want to use software, such as QuickBooks to log your income and expenses. Software can make things much easier to keep up with, so consider a software package that's easy to use and helpful.

» **Keep up with every expense.** Every single thing you spend on your business is a business expense that you need to log and have receipts for. Did you pay a website hosting fee? That's a business expense. Did you buy a new ND filter set for your drone? That's a business expense. Log everything. You'll be surprised how small items add up over time.

» **Be careful with advertising.** Advertising expenses can eat virtually all your profit. Though advertising is necessary, you'll need to find the right balance so that you don't spend every dollar you make trying to find new customers.

» **Consider separate bank accounts for your business.** You'll probably need a business checking or savings account to cash checks made out to your business. Check with your bank for details.

» **Accept online payments.** Venmo and PayPal are popular online payment options, and your website platform may even accept them as well. Customers will often want to pay electronically, so make it easy for them to pay you!

Considering drone insurance

Even the best drone pilots have accidents, so it's a good idea to have liability insurance in case you crash into someone's personal property while you're in the middle of a drone piloting gig.

Start with your home or auto insurance company and see whether they have any affordable business insurance for you. Remember that insurance is an expense, so you'll have to keep the level of insurance in mind as you think about your expense and income.

TIP

Also, some companies specifically provide drone insurance, sometimes on a job-by-job basis. This type of insurance may be more affordable (and some jobs will require it). Some good places to get started checking out drone insurance are https://www.skywatch.ai/ and https://www.airmodo.com/.

Paying your dreaded taxes

Death and taxes are certainties; you know that! Drone piloting is no exception. Because you have a business, you'll need to pay both state and federal income tax, and in some cases, you may need to pay local sales tax as well.

Generally, you'll pay these taxes on a quarterly basis, and how you pay them depends on the state you live in. The best way to get started is with an accountant. The accountant can help you get things set up so that you can make sure you pay your taxes on time.

REMEMBER

Many small businesses get into financial trouble because they neglect to pay their taxes. Get organized, pay your taxes, and avoid all those fees. Organization matters!

6

The Part of Tens

Chapter **20**

Ten Part 107 Test-Day Tips

When you think about taking the Part 107 exam, you may feel like you're sitting in Mrs. London's high school English class. You may think, "I didn't do so well on tests back then, and it's been awhile since I've had to take a multiple choice test!"

Well, misery loves company! You're not alone; most people do not enjoy taking tests. Naturally, you need to study thoroughly before you take the Part 107 exam, but you can also do a few things on the test day to give yourself an edge. Put the following ten test day tips to work.

Eat, Sleep, Repeat

Most people understand the principle of eating and resting well before you take a test. However, you may not realize the need to establish this pattern a few days before the test, not just in the hours before it.

So after you schedule your test, force yourself to get into a good sleep pattern at least a few days beforehand. Aim for seven to eight hours of sleep each night and be sure to go to bed and arise the next morning at the same times each day.

Also, strive to eat nutritious, well-balanced meals and drink plenty of water each day. You want to be well fed and well hydrated. Avoid alcohol like the plague before the test day. Also, don't overeat. Now is not the time to enter the local chicken wing–eating contest. (Save that as a celebration after you pass the test.)

Your body works best when you follow a routine. When you eat the same types of meals every day, get the same amount of sleep, and stay hydrated, all on the same basic routine, your body will be well rested and energized and your brain will work better. That's what you want!

TIP

One more aspect of getting ready for test day is to think carefully about what time of day works best for you to schedule the test. Most people perform better on tests during the morning hours. As the day progresses, your brain and eyes get tired, so a morning test is often the best option.

Buy a Magnifying Glass and Cheap Calculator

Sectional charts are crowded with information, with some of it appearing in very small print. You can pick up a cheap magnifying glass and take that with you into the testing center. Also, a cheap, basic calculator is a good idea to have on hand as well. You don't need either item to be complicated or expensive. You should be able to pick up these items for few bucks, but they can help, so spend a little coffee money on them!

REMEMBER

You can't take your phone into the testing center, so you'll need a cheap pocket calculator instead. You may not need either of these items, but it's better to have them and not need them than to need them and not have them!

Make Notes Before You Start

The testing center will give you some scratch paper and a pencil. When you sit down to take the test, you click a Start button on the computer when you're ready to get under way, which also starts your testing time. Before you click Start, grab the test paper and a pencil and make some notes.

Write down details that you have trouble remembering or may get confused about, such as airspace details, radio frequencies, and other easy-to-forget items. Basically, make yourself a cheat sheet that is legal to use on the test because you're creating it there. When you're taking the test, you can quickly refer to your notes.

Slow Down

When people are anxious, they tend to move and think more quickly. These are natural bodily reactions to anxiety, but they can cause problems when you take the test. If you're thinking too quickly, you'll make careless mistakes.

So consciously work on slowing down. Take some slow, steady breaths and force yourself to read more slowly. Read each test question twice to make sure you understand what the question is actually asking you (not what your first glance suggests that it's asking you).

REMEMBER

You can't make yourself not be anxious, but you can strive to mitigate it by using a simple breathing exercise, for example.

Increase Your Odds

Each Part 107 test question (of which there are 60), will give you three answer options: A, B, or C. To increase your odds of finding the correct answer, try the following tips:

>> **Choose the best answer.** People tend to think of a test as having correct and incorrect answers, but that's not entirely true. On a test like the Part 107, you want to look for the *best* answer, not the correct answer. That's because two answer options may sometimes be similar and both could be correct in different situations, so of the two, you need to decide which answer is the best. Go back and think about what the question is asking you, and then determine the best answer to the question. Get in the habit of looking for the best answer instead of the correct answer. This will help you get some of the questions right!

>> **Use the process of elimination.** For almost all the questions, one answer option will be completely wrong. Instead of first looking for the best answer, ask yourself, "Which answer option is wrong?" When you identify the wrong answer, you can eliminate it, thereby leaving only two options and giving yourself a 50/50 shot at guessing the correct answer if you're not sure.

>> **Double-check the answer you choose.** Read the answer option, and then read it again. Sometimes the answer may not actually say what you initially think it says. Slow down and read it twice!

Read at "Face Value"

A big problem many people unwittingly create with standardized tests is to read their own life experiences into the test questions. For example, if a question asks you about a drone's battery life during a flight, you may read the answer options and think, "Well, it depends. If the wind is blowing a lot, that will degrade battery power." But here's the thing: The test question didn't tell you the wind was blowing. You're reading that information into the question, which can cause you to answer incorrectly.

You have to make it a point to read the test question at face value — even if the question doesn't seem reasonable or reflective of a real-world situation. Don't add your own interpretation to it. Just answer the question based only on the information given to you.

Mark Tricky Questions

REMEMBER

On test day, you'll encounter questions that, at first, you believe you answered correctly, but then you're not sure. If you're not careful, you can waste precious time staring at the question and thinking about it. You have two hours to answer 60 questions, which gives you two minutes per question. So you can't spend a lot of time on each question.

Here's the best way to ensure that you get through all the questions. Each question will give you the option to *mark* it. Mark any question whose answer you're not sure of, and at the end of the test, you can go back and review the questions you marked. Then you can use what time you have left to review the questions you marked. If you don't have time to get back to all of them, that's okay. You're not penalized for marking a question.

There's one caveat, though. Marking questions is a good strategy, but you also want to be careful using it. In almost all cases, your first, gut-level instinct is correct. If you review a question too much, you start overreading the question and move beyond a face-value look. That can cause you to miss the answer.

So feel free to review, but use extreme caution when you change an answer. Your first instinct tends to be correct.

Quickly Guess at Questions You Don't Know

Some questions will be tough. Even if you work on them and believe you have the right answer, those questions are good to mark and review at the end of the test, as suggested in the preceding section. However, you'll also see some questions whose answers you probably won't know at all. Some of them may not even appear to make sense!

If you read a question and immediately think, "I have no idea,", then do this: Eliminate an answer option if you can. Eliminating an option that you're pretty sure is wrong will at least get you to a 50/50 chance of guessing the answer correctly. Then just guess. Don't mark the question if you have no idea. Just take a guess and move on quickly. Save time to work on questions you may be able to answer correctly.

Dress for Success

You may be thinking, "Do I need to wear a suit to this test?" No, of course not, but your clothes may matter. Some testing centers are cold, and if you're cold, you're uncomfortable and easily distracted. So, do this: Bring a jacket or hoodie just in case. I took my Part 107 test in August, so I wore shorts and a T-shirt. But I also took a hoodie because I hate being cold! If you don't need the jacket or hoodie, no worries; but if you do, you'll be glad you have it.

Otherwise, just dress comfortably. You're not trying to impress anyone at the testing center.

Know the Test Details

The Part 107 exam contains 60 multiple-choice questions. You have to score 70 percent to pass, which means that you need to answer 42 questions correctly. The 60 questions you get are randomly selected from a larger bank of questions, so no two tests are the same.

You have two hours to take the test. Doing the math, that means you can spend up to two minutes answering each question. In reality, you can answer some questions in a handful of seconds; others take longer. But you'll have plenty of time. The questions tend to be short and to the point, so you won't have to spend time reading a bunch of convoluted questions — which is great news!

Remember That the World Won't End

Here's the last tip and one of the most important: If you don't pass the test the first time, the world won't end; your life won't fall apart; you won't blink out of existence. The Part 107 test is not a life and death situation, so don't let your anxiety make it seem like one.

Repeatedly tell yourself the following: "If I fail the test, I'll just regroup and take it again. Nothing bad is going to happen."

When you make these statements your mantra, you keep things in perspective, which in turn reduces stress. People who are less stressed think more clearly. And, people who think more clearly usually pass.

Now, read the final two chapters of this book, go tackle the test, and best of luck to you!

Chapter **21**

Ten Drone Piloting Myths

Drone piloting is a complex subject that involves working with a lot of information. After all, that's why you're reading this book! However, most complex subjects also tend to generate myths, or at least misunderstanding. For example, I see many statements on drone piloting social media groups that just aren't true. People believe they understand something about the topic and then confidently discuss it on the internet as though they know the truth, when they're actually just propagating misinformation.

This chapter takes aim at ten common myths that circulate about drone piloting. Although I cover some of the topics you read about here in more detail throughout the book, this chapter helps you keep these myths from tripping you up.

You Don't Need Certification or Registration If Your Drone Is Small

People often confuse registration and certification, and the FAA convolutes the two in some ways. Drones are registered; people are certified. But the two concepts work together. Here are the actual rules to remember:

» If the drone weighs under 250 grams, you are not required to register it with the FAA if the drone is used only for recreational purposes. All drones weighing over 250 grams must be registered at https://faadronezone-access.faa.gov whether you're a certified pilot or not.

>> If you fly your drone only for personal enjoyment, you are not required to have a Part 107 certification, regardless of the drone's weight (though you do have to register any drone over 250 grams regardless of how you use it). *Personal enjoyment* means that you fly your drone for fun and solely as a hobby. You can take photos and videos and use them on your personal social media if you like, as long as your social media isn't monetized.

>> If you have your Part 107 certification, you must register any drone you use for commercial purposes at https://faadronezone-access.faa.gov. This requirement includes drones that weigh under 250 grams.

>> If you want to use your drone for any commercial purpose, you must have your Part 107 certification. This requirement applies in all the following ways:

- Any nonprofit work. The FAA considers work for nonprofit organizations "commercial."

- Whether you are paid or not.

- Any drone you use, even one that weighs less than 250 grams. See Chapter 5 for details.

So just to be crystal clear, you *do* have to register a drone that weighs in under 250 grams if you are a certified UAS pilot and you use that drone for commercial purposes.

Practically, here's the easy way: If you have your Part 107 certification, register every drone you have no matter how you'll use it. That way, if you decide to use a drone that weighs under 250 grams for commercial purposes, you're legal to do so. Even if you have a sub-250-gram drone for recreational use, you never know when you might need it for something commercial. Registration is quick and inexpensive, so just register all the drones you own; then you don't have to worry about it.

Crewed Aircraft Do Not Fly under 400 Feet

The assertion in the heading for this section is not true! Although larger aircraft do not fly under 400 feet except for takeoff and landing, other aircraft may fly lower than 400 feet. For example, a helicopter may fly under 400 feet for a variety of reasons, depending on its mission. Small planes used in agriculture often fly under 400 feet for crop dusting and such. The point is that some drone pilots think that as long as they fly under 400 feet, they don't have to worry about other aircraft. This simply isn't the case.

REMEMBER

The FAA expects you to maintain situational awareness and always yield right of way to aircraft with human beings onboard. That's why you need to be able to see your drone when you fly. You may need to make quick decisions to avoid a collision. No matter where you're flying, always be on the lookout for other aircraft and immediately yield right of way.

You Don't Have to Worry About Airspace as Long as You Fly under 400 Feet

Different airspaces have different rules. That's why the FAA expects you to know about airspace (see Chapter 9). In many cases, you'll need permission to fly in restricted airspace regardless of the height you plan to fly. In some cases, you'll receive permission to fly in restricted airspace but with a ceiling of perhaps only 200 feet.

The FAA partners with app companies so that you can check the airspace you're in and attempt to gain permission if necessary. So the best practice is to always check the airspace in an area where you plan on flying. You'll need to know what airspace you're in, whether it has any specific restrictions, and what permission you might need.

TIP

You can check out the different B4UFLY apps available at https://www.faa.gov/uas/getting_started/b4ufly. Download one of these apps and put it use!

You Don't Need Line of Sight If Your Drone Has a Good Camera

The FAA requires you to maintain visual line of sight (VLOS) when you fly your drone. This means you must be able to physically see your drone without the aid of binoculars or other visual aid, and you must maintain situational awareness of what's happening in the sky around your drone.

Note that first-person view (FPV) flights do not satisfy the FAA's VLOS rules. When you fly using FPV, you wear FPV goggles that have a built-in screen. If you're using FPV, you need another person to serve as a visual observer to help you maintain VLOS of your drone as you use FPV.

The drone's camera system has nothing to do with VLOS. The FAA requires you to physically keep your eyes on your drone as you fly and to know what's going on around the drone as well.

You Don't Have to Report an Accident to the FAA

If you have an accident with your drone, you are required to report that accident to the FAA if

>> The drone accident causes more than $500 in damage to someone's property.

>> The drone strikes a person and a loss of consciousness occurs.

>> The accident causes a serious injury. *Serious injury* means that medical attention is needed. If a drone propeller blade scratches a person's arm, you don't need to report that injury.

REMEMBER

Your obligation to report has nothing to do with the injured person's response or actions. You must report the accident if one of the three preceding conditions is met. You can report accidents to the FAA through your DroneZone account.

You Don't Have to Worry About MTRs

Military Training Routes (MTRs) are found on sectional charts. Basically, they are routes in the National Airspace System where military training flights may occur under 10,000 feet. Often, these training sessions include fast-flying aircraft.

MTRs are operated by the local military base. If a training is scheduled to occur on an MTR, the base will send out a Notice to Air Missions (NOTAM) at least two hours before that training stating when the training will occur. (See Chapter 10 for more information about NOTAMS.)

Although MTRs are safe areas for drone flights under 400 feet, you still need to be aware of what may be occurring overhead as you fly. Military training often involves fast-flying aircraft, and as always, drone pilots must yield right of way. Although you can fly in an MTR, you want to use caution and be situationally aware.

Elevation Doesn't Impact Drone Battery Life

There's often such a focus on the 400-foot maximum flight height that new drone pilots fail to consider elevation. Keep in mind that 400 feet means 400 feet above ground level (AGL). Practically, this means that if you take off from a beach location, you can fly 400 feet high, but if you take off from the top of a big mountain, you can also fly 400 feet from that takeoff point. In both cases, you're flying 400 feet AGL, but the elevation is quite different in these two cases.

Here's the thing: Elevation can affect your drone's battery life. Flying in thin air (high elevation) requires the drone to work harder to maintain lift. Conversely, flying in a humid area (low elevation, thick air) requires the drone to work harder to fly. In both cases, the elevation will impact and likely reduce the amount of time you have to fly because the battery won't last as long. In a nutshell, if the drone has to work harder to fly, the battery life will be shorter. Keep battery life in mind as you fly at different elevations.

After You Learn How to Fly, You Don't Need RTH

Most prosumer and professional drones have a Return to Home (RTH) feature. This feature enables the drone to know where it took off from using GPS coordinates and to return to that location on its own. Although you may not need to use RTH because you can fly and land the drone yourself, many drone pilots develop a flippant attitude about RTH and fail to realize its importance.

The issue of "lost link" is what makes RTH a really important strategy. The term *lost link* refers to the loss of communication between a drone in flight and the remote controller, a problem that arises for almost all drone pilots at one point or another. (See Chapter 15 for details.) Using RTH, the drone takes over, returns to the landing site, and lands without instructions from the controller. This is a critical feature that can help you avoid crashes, damages, and potential injury.

REMEMBER

As a safety precaution, always give the drone and controller time to update the home point before you fly. In the case of a lost link problem, RTH can save the day!

The Built-In Drone Lights Are All You Need to Fly at Night

Most drones have lights built onto the wings. However, those lights typically are not enough to satisfy the FAA requirements for night flights. If you want to fly your drone at night, you need to add anti-collision lights that are visible from three statute miles away.

Numerous anti-collision light kits and products are available, so you can easily add these lights if you need them.

Drone Piloting Jobs Don't Require State Sales Tax

You may hear that you don't need to pay state sales tax when you are paid for a piloting job. Don't count on that idea. For example, a neighbor hires you to photograph a piece of property and pays you after you take the shots. If you have an established business, such as a limited liability corporation (LLC), you probably have to pay state sales tax on the sale.

The rules for paying sales tax depend on your business and the state you live in, so you need to check out these requirements and pay the sales tax if required. Often the penalties for not paying this tax are high, so don't want those!

Chapter **22**

Ten Common Drone Piloting Problems You Can Avoid

M urphy's Law states, "Anything that can go wrong will go wrong." That adage feels really negative, but in many cases it proves true, and drone piloting is no exception. Things go wrong; equipment and software do fail from time to time. Often, however, what went wrong could have been avoided!

With that thought in mind, this chapter presents ten common problems that drone pilots experience and how you can generally avoid them.

Software Glitches and Errors

Software is made by people, and people aren't perfect, so your drone's software isn't perfect, either. Sometimes the software contains bugs that can even affect the safety of your flights.

But here's the truth: Most pilots who have software problems do so because they do not keep the software updated. Drone manufacturers release regular software updates to fix problems and roll out new features. You need these updates!

REMEMBER

At least every other week or so, check your drone remote controller (RC) for updates that may need to be applied to the drone or the RC itself (or both). In most cases, updates take only a few minutes, and those few minutes are well worth your time. See your drone's documentation for details about checking for updates; also see Chapter 2 for more information about updates.

Shaky Video and Blurry Images

If the video on your RC looks shaky and the images you capture are blurred, you may have a problem with the camera gimbal, or even the camera lens. But before you contact tech support, first do a careful inspection of the propellers.

As a habit, you should check the propellers for any damage before every flight (see Chapter 4 for details), but some drone models are particularly sensitive to a single propeller that is even slightly bent or chipped. This slight damage can cause a lot of vibration as your drone flies, resulting in poor video and image quality. You may even miss the damage with a quick look, so slow down and look closely at each of the propellers. Also helpful is to run your fingers down them to check for any warping or bends. If you find any flaws, replace the faulty propeller and try your video and photo shoot again. If that doesn't resolve the problem, it's time to contact tech support.

Flying in the Wrong Airspace

Flying your drone where you shouldn't happens commonly, and it's completely avoidable! Sometimes pilots *believe* they are in class G airspace without any restrictions and take off, only to discover that they needed LAANC approval or can't fly there at all.

REMEMBER

Sometimes airport airspace is bigger than you think, or you may be near a national park or other government land with restrictions. Whatever the reason may be, the point is this: Don't *believe* you're okay; check first to make sure. It's easy and very fast to check the airspace using one of the FAA-approved apps (see Chapter 2 for more information). Although you may get away with flying in controlled airspace, you may also incur a stiff fine if caught. Also, controlled airspace is controlled for a reason; you don't want to be the cause of an accident. So play it safe and check it out first!

Connectivity Problems with the Drone and RC

Possible reasons abound for connectivity problems with the drone and RC. Maybe you can't get a home point established, or the RC never gives the go-ahead to take off because it didn't connect to enough GPS satellites. Although some connectivity issues require help from tech support, check one thing first: whether you've used the correct sequence to power on your devices.

Always turn on the drone before turning on the RC. The RC needs to first detect the drone's presence, and going out of order can keep them from connecting properly.

So if you're having problems, shut down the drone and RC, and then try powering them up again in sequence (drone first, then the RC) to see if doing so resolves the problem.

TIP

If the problem persists, start with the drone manufacturer's website for troubleshooting steps. Although the internet holds a wealth of information, it also holds a welter of disinformation, and people post all kinds of troubleshooting advice that just isn't accurate. Save some time and start with the manufacturer's website.

Overestimating Battery Life

No drone's battery lasts as long as its user wants it to. However, if your drone typically has a battery flight time of, say, 34 minutes, that may change depending on conditions.

REMEMBER

Keep in mind that wind, air density, temperature, and other atmospheric conditions affect battery life. The wind can especially impact battery life, and even with the Return to Home (RTH) feature, the drone can run out of battery power before getting home in very windy conditions.

My greatest close call happened while flying on a windy day and overestimating how much battery life remained. I barely got my drone back and landed. Always think carefully about battery life and what conditions are impacting it. You'll be glad you did!

Losing VLOS

Drone advertisements may say things like, "Nine-mile range." Well, unless you have really good eyesight, you're not supposed to fly that far. As noted in Chapter 6, the FAA requires that you maintain visual line of sight (VLOS), meaning you must be able to see your drone in the air without the aid of binoculars or other devices.

VLOS doesn't always seem very practical depending on what you're doing. I get it. However, the fact remains that most drone crashes and accidents happen when the pilot can't see the aircraft. You lose situational awareness and end up in a circumstance you can't see coming when you're restricted to just the RC screen.

So follow the rules and keep your drone in your physical sight. Although it's tempting to fly far, far away, it takes only one crash to regret it.

Flights Seem Tilted or Off Balance

If you're flying your drone and notice that the RC screen image seems a bit off balance or tilted, or if the drone behaves a bit erratically, land it! Then check the drone's manual for instructions about calibrating the compass and the gimbal.

For most drones, the calibration process is quick and easy, and it helps the drone understand direction if something is off with the compass. If calibration doesn't resolve the problem, it's time to reach out to tech support.

Poor GPS Connectivity

Your drone uses GPS to "see" where it is in the world. The more GPS satellites your RC connects to, the better. But what if after waiting patiently, you can connect to only a few satellites?

REMEMBER

Here's the thing: Sometimes structures, the landscape, and even underground cables can interfere with GPS signals. If you're in a location where you can't seem to get a good connection to GPS satellites, just try moving to another location nearby. This simple step may resolve the problem. GPS connectivity isn't necessarily a problem with your RC; it may have more to do with your location. So when in doubt, just move.

Overdependance on Obstacle Avoidance

Drone sensors and obstacle-avoidance features are simply great. They can help you avoid all kinds of potential pitfalls and crashes. But no sensor system is perfect. You need both the sensor system and your eyes to see what is going on around you. Always look at what your drone is flying around and make safe flight decisions based on what you see.

When pilots become overdependent on obstacle-avoidance systems, accidents happen. The drone may not see something small in its path, such as a small tree branch, and run directly into it.

So use obstacle avoidance. It can help you. But don't forget to use your eyes.

You

I'm sorry to say that one of the biggest drone piloting problems is, well . . . all of us drone pilots! Drone piloting is a technical game, but it's also a mental game. Many different factors can interfere with sound judgment when you have to make quick decisions. Low blood sugar, lack of sleep, tired eyes, stress, alcohol, and medications are just a few of the impediments to a pilot's performance.

The key? Self-awareness. Don't be overconfident. Be honest enough to take a good look at yourself and say, "I'm just not in the best condition to be flying right now." That statement shows wisdom. A good pilot is always a safe pilot, so if you're not on top of your game for any reason, delay the flight. There's always another day to fly!

Index

A

abort, knowing when to, 255

above ground level (AGL), 100–102, 325

accessories, tips for, 24

accidents, reporting to FAA, 117–118, 324

accuracy, of Global Positioning System (GPS), 34

acronyms

  about, 163–164, 165

  Military Training Route (MTR), 167–170, 324–325

  Notice to Air Missions (NOTAM), 165–166, 324–325

  Temporary Flight Restriction (TFR), 166–167, 168, 169

  Victor Airway, 170–171

  Visual Flight Record (VFR), 172

Active Track option, 60

active tracking, shooting with, 59–60

ADM (Aeronautical Decision Making), 249

Adobe Lightroom, 275

Adobe Premiere Pro, 289

advection fog, 228

advertising, 308–309, 311

AE (Automatic Exposure) mode, 262

aerial eye-level angle, 270–271

aerial-down angle, 270

AeroLion BlackLion 078 drone, 299, 300

Aeronautical Decision Making (ADM), 249

aftermarket propellers, 68

age

  Part 107 exam and, 87

  of propellers, 67

AgEagle's eBee X drone, 114

AGL (above ground level), 100–102, 325

agriculture, drone piloting for, 9, 296–297

aiming point, as a runway marking, 214

air masses

  about, 222

  cloud types, 222–223

  density altitude, 224

  stable air, 224

  unstable air, 224

Air Traffic Control (ATC), 119, 124, 144, 145, 149, 210

air travel, with drones, 31

aircraft

  crewed, 322–323

  how they land, 214–218

  in PAVE model, 252

Airman Knowledge Testing Supplement for Sport Pilot, Recreational Pilot, Remote Pilot, and Private Pilot, 164

airports

  flying drones around, 207–218

  nontowered, 210–211

  operations, on Part 107 exam, 91–92

  questions about, 192–197

  right of way, 218

  runways, 211–214

  towered, 210–211

  traffic at, 211–218

airspace

  about, 323

  flying in the wrong, 328–329

  obtaining authorization, 119

  obtaining authorization for night flights, 124

  on Part 107 exam, 87–88

  questions about, 192–197

alcohol, 76, 121, 252

Alert area, 159, 160

Allstate, 25, 26

altocumulus clouds, 223

altostratus clouds, 223

AMEs (Aviation Medical Examiners), 121

anatomy, of human eyes, 129–130

angles, for photos, 269–273

anti-authority, 253

anti-collision lighting, for night flights, 125

aperture, 264–265

Apple Final Cut Pro, 289

apps, 35–36

Arctic air masses, 222

arranging video clips, 290

ascend movement, 282–283

Asurion, 25

ATC (Air Traffic Control), 119, 124, 144, 145, 149, 210

ATIS (Automatic Terminal Information Service), 174–175

attention to detail, as a characteristic of pilots, 11

Autel Evo Max drone, 22, 23, 301

Autel Robotics, 21–22

authorized airspace, 75

Auto mode, 261

autokinesis, 133–134

automated video footage, 62

Automatic Exposure (AE) mode, 262

Automatic Terminal Information Service (ATIS), 174–175

Available card storage, as onscreen feature, 34

Aviation Medical Examiners (AMEs), 121

Aviation Weather Center, 229

avoiding distractions, 237

B

B4UFLY app
  about, 35–36, 144, 149, 154, 161, 170
  in pre-flight checks/procedures, 39, 47
  website, 323

backpacks, for drones, 43

bank accounts, for businesses, 311

bank angle, 246

base leg, 215–216

basic maneuvers, practicing, 47–54

basic warranty, 25

batteries
  for drones, 31
  elevation and life of, 325
  emergencies with, 235
  limitations of with night flights, 128
  overestimating life of, 329–330

in post-flight check, 42, 43
  in pre-flight checks/procedures, 39

Battery remaining, as onscreen feature, 33

beach photos, Part 107 certification and, 83

best practices, for night flights, 136–139

billboards, advertising on, 309

birds, emergencies with, 235–236

bird's-eye movement, 285

Brake feature, as an obstacle-avoidance behavior, 36–37

Brinc Drones Lemur 2 drone, 298

broadcast module, Remote ID, 107–108

built-in lights, 326

built-in maneuvers, 58–59

built-in software, 289

bulb air blower, 69–70

Burst mode, 262

BWINE F7 series, 20

Bypass feature, as an obstacle-avoidance behavior, 37–38

C

calculator, 316

calibrating drones, 70–71

call sign, on METAR, 229

calmness, as a characteristic of pilots, 11

camera gimbals, 17–18, 24

cameras
  cleaning, 68–70
  limitations of with night flights, 126–127
  troubleshooting issues with, 74

capability, Part 107 exam and, 87

careers, for drone pilots, 302–304

carelessness, emergencies with, 236

cases, for drones, 43

categories, for drones, 19–24, 115–116

causes, of emergencies, 234–236

ceiling, 145

center of gravity, 247–248

centerline, as a runway marking, 214

certified remote pilot, 30

certified remote pilots, TRUST certificates and, 82–83

changing propellers, 66, 68

charging LiPo batteries, 72–73

chart supplements, reading, 208–209

Cheat Sheet (website), 3

checking for updates, 30–31

checklists, Crew Resource Management (CRM) and, 93

Choose an action, in DECIDE model, 250

choosing first flight locations, 38–39

circle maneuver, 50, 51

cirrocumulus clouds, 223

cirrostratus clouds, 223

cirrus clouds, 223

Class A airspace, 144–145

Class B airspace, 145–148

Class C airspace, 149–151

Class D airspace, 151–153

Class E airspace, 153–157

Class G airspace, 29, 157

cleaning

  cameras, 68–70

  drone body, 68–69

  drone camera, 69–70

  drones, 68–70

  gimbal, 69–70

  propellers, 68–69

cloud types, 222–223

collision-avoidance sensors, 287

color

  eyes and, 131

  improving in photos, 275

  in videos, 291

color grading, 291

commercial photos, Part 107 certification and, 85

Common Traffic Advisory Frequency (CTAF), 173–174

communications

  Crew Resource Management (CRM) and, 93

  skills, as a characteristic of pilots, 11

Communications section, on charts, 208

compass directions/degrees, 211–214

complex maneuvers, 54–58

complying with Remote Identification (Remote ID), 106–109

composition, applying, 268–269

cones, in eyes, 130–132

confidence, growing, 63

connectivity

  GPS, 330–331

  issues with, 329

  troubleshooting, 74

construction, drone piloting for, 9, 299–300

consumer drones, 20–21

Continental air masses, 222

control tower (CT), 174–175

Control tower, on charts, 209

controlled airspace, 143

convection, 222

convection currents, 225

conversations, while flying, 76

coordinates, GPS, 182

cornea, 129

Corpus Christi International Airport (CRP), 149

cost, 24

crash safety, 36–38, 40

crew members, 92

Crew Resource Management (CRM), 92–93

critical angle of attack, for fixed wing drones, 248–249

CRM (Crew Resource Management), 92–93

cropping photos, 274

crosswind leg, 215–216

CRP (Corpus Christi International Airport), 149

CT (control tower), 174–175

CTAF (Common Traffic Advisory Frequency), 173–174

cumulonimbus clouds, 223

cumulus clouds, 222

Current camera selections, as onscreen feature, 34

D

D05 (Garrison Municipal Airport), 155–157

Dallas Fort Worth Airport (DFW), 165–166

damage, 42, 234

dark conditions, adapting to, 132

DaVinci Resolve, 289

day, on METAR, 229

DBA (doing business as), 305

DECIDE model, 250–251

Declaration of Compliance (DOC), 114, 115

delivery drones, 9, 301

density altitude, 224

descend movement, 282–283

Detect the problem, in DECIDE model, 250

dew point, 228, 230

DFW (Dallas Fort Worth Airport), 165–166

DI Inspire, 22, 23

digital negative (DNG) file type, 267

direction, runways, landing, and, 216–218

displaced threshold, as a runway marking, 214

distractions, avoiding, 237

DJI drones, 21–22

DJI MasterShots, using, 61–62

DJI Matrice 3DR Solo drone, 298

DJI Matrice 350 RTK drone, 298

DJI Mavic 3 Pro drone, 22, 23, 126, 301

DJI Mini series, 20

DJI RC Pro drone, 22, 23

DNG (digital negative) file type, 267

Do the necessary steps, in DECIDE model, 251

DOC (Declaration of Compliance), 114, 115

doing business as (DBA), 305

domain names, 307

downdrafts, 227

downwind leg, 215–216

drag, 12, 244

drone body, cleaning, 68–69

drone camera, cleaning, 69–70

Drone Deploy, 295

Drone Log Book, 118

Drone Pilot Ground School, 94

drone piloting

  about, 7–8, 293

  advertising your business, 308–309

  for agriculture, 296–297

  areas of interest, 303–304

  in business, 294–302

  for construction and inspections, 299–300

  current job, 302–303

  defining your business, 305

  drone insurance, 311–312

  establishing internet presence, 307–308

  for film and photography, 301

  finding a career, 302–304

  freelance business, 304–312

  getting paid, 309–310

  as a hobby, 8

  juggling income and expenses, 311

  for law enforcement, 297–298

  making first contact, 306

  myths, 321–326

  paying taxes, 312

  problems to avoid, 327–331

  as a profession, 8–10

  for real estate and mapping, 294–296

  remote, 10–11

  for search and rescue, 298–299

  for security, 301–302

  setting up your business, 305–306

  for shipping and logistics, 300–301

  skill development, 304

  state sales tax and, 326

drones

  air travel with, 31

  backpacks for, 43

  calibrating, 70–71

  camera gimbals, 17–18

  cases for, 43

  categories for, 19–24, 115–116

  cleaning, 68–70

  consumer, 20–21

  flying around airports, 207–218

  how they work, 12–18

  industry-specific, 24

  insurance, 25–26

  limitations of with night flights, 126–128

  lost, 241

  professional, 22–23

  prosumer, 21–22

  registering, 28–30

  remote controllers (RCs) for, 15–17

  reviews for, 23

shopping for, 18–24
storing, 73
toy, 19–20
warranties, 25–26
weight maximum, 18–19
weight of, 80–81
DroneUp, 301
dronie, 286
drugs, 76, 121

E

Easy Aerial SAMS, 302
editing
photos, 274–275
video, 288–292
effects, adding to videos, 292
elevation
about, 183
battery life and, 325
maximum elevation figures (MEFs), 184–185
natural maximums, 184
obstructions, 185–186
questions about, 197–199
towers, 185–186
emergency scenes/personnel, 75
Emotions, in IMSAFE model, 252
Environment, in PAVE model, 252
environmental effects, troubleshooting, 74
Epidemic Sound, 292
Estimate the need to react, in DECIDE model, 250
Evaluate your action, in DECIDE model, 251
Evo Lite drone, 21–22
Evo Nano drones, 21–22
exams, Part 107, 86–97
exceptions, to maximum height, 102–103
EXO Drones, 21–22
expenses, juggling with income, 311
exporting video, 292
exposure
about, 262–263
aperture, 264–265
improving in photos, 275

International Organization for Standardization (ISO), 266
shutter speed, 263–264
in videos, 291
extended warranty, 25
External pressure, in PAVE model, 252
eyesight limitations, for night flights, 128–132

F

FAA. *See also* flight regulations
about, 18
Airman, Knowledge Testing Supplement for Sport Pilot, Recreational Pilot, Remote Pilot, and Private Pilot, 164
registering drones with, 28–30
regulations for night flights, 124–125
regulations from, 74–75
reporting accidents to, 117–118, 324
website, 30
FAA Drone Zone, 29, 81, 118
FAA-Recognized Identification Area (FRIA), 108–109
family reunion photos, Part 107 certification and, 83–84
fascination, 133
Fatigue, in IMSAFE model, 252
file types, for photos, 266–267
film, drone piloting for, 9, 301
final leg, 215–216
firmware, 30–31, 66
first-person view (FPV), 32, 104, 297, 323–324
fixed-wing drones, critical angle of attack for, 248–249
FJI Terra, 295
flashcards, for Part 107 exam, 95
flicker vertigo, 134
flight emergency procedures
about, 233
common causes of emergencies, 234–236
lost-link procedures, 238–242
practicing foundational emergency responses, 237–238
Flight height, distance, and map, as onscreen feature, 34

Flight mode, as onscreen feature, 33

flight principles, 46–47

flight regulations

about, 99, 111

above ground level (AGL), 100–102

complying with remote identification, 106–109

effects of alcohol/drugs, 121

flight-readiness issues, 122

flying from moving vehicles, 117

flying over large groups of people, 116–117

flying over people, 113–116

flying over property, 112–113

following, 99–109, 111–122

logging flights and maintenance, 118–119

maximum height, 100–103

mean sea level (MSL), 100–102

National parks, 121

obtaining airspace authorization, 119

obtaining LAANC authorization, 120

on Part 107 exam, 88–89

reporting accidents to FAA, 117–118

speed, 103–104

during twilight, 105–106

visual line of sight (VLOS), 104–105

weather visibility, 103–104

"flight time," tips for, 24

flights, logging, 118–119

floor, 145

fly in movement, 285–286

flyaway, 25

flying

about, 27, 45

B4UFLY app, 35–36

built-in maneuvers, 58–59

checking for updates, 30–31

choosing locations for first flights, 38–39

combining skills, 58

crash safety features, 36–38

drones around airports, 207–218

frequency of, 63

geofencing, 38

Global Positioning System (GPS), 34–35

growing skills and confidence, 63

landing for first time, 40–42

at night, 123–139

over moving vehicles at night, 124

over people at night, 124

post-flight checks, 42–43

practicing basic maneuvers, 47–54

practicing complex maneuvers, 54–58

pre-flight checks/procedures, 39–40

principles of, 46–47

registering drones, 28–30

remote controllers (RCs), 32–34

shooting with active tracking, 59–60

taking off for first time, 40–42

troubleshooting, 42

using JI MasterShots, 61–62

Flytrex, 301

focusing problems, during night flights, 127

fog, 227, 228–229

following flight regulations, 99–109, 111–122

force equilibrium, 244–245

foundational safety practices, 74–76

fovea area, 131

FPV (first-person view), 32, 104, 297, 323–324

freelance business, setting up, 304–312

freezing fog, 228

FRIA (FAA-Recognized Identification Area), 108–109

FTP (pre-planned flight termination points), for list-link emergencies, 241

G

Gainesville Municipal Airport (GLE), 210, 211

Garrison Municipal Airport (D05), 155–157

Geico, 26

geofencing, 38

getting paid, 309–310

gimbal rotation maneuver, 56–57

gimbals

about, 17–18, 56

cleaning, 69–70

tips for, 24

GLE (Gainesville Municipal Airport), 210, 211
Global Positioning System (GPS)
  about, 34–35
  accuracy of, 34
  connectivity, 330–331
  coordinates for, 182
  for lost drones, 241
goals, 24
GPS (Global Positioning System)
  about, 34–35
  accuracy of, 34
  connectivity, 330–331
  coordinates for, 182
  for lost drones, 241
GPS Connections, as onscreen feature, 33
graininess, during night flights, 126, 127
gravity, center of, 247–248
green lights, on airplanes, 135–136
ground level, Class E airspace starting at, 154–155
guessing, during Part 107 exam, 319

H

hail, tips for, 232
handling lost-link emergencies, 239–242
height, maximum, 100–103
help, getting, 74
hobby, drone piloting as a, 8
Holy Stone HS600 drone, 20, 21
humidity, 227–228, 232
Hyllo AG-230 drone, 297

I

ice fog, 228
icons, explained, 3
Identify a solution, in DECIDE model, 251
IFR (instrument flight rules), 144, 168
illness, in IMSAFE model, 251
importing video clips, 289
impulsivity, 253–254
IMSAFE model, 251–252
IMU (Inertial Measurement Unit), 70

income, juggling with expenses, 311
incorrect thinking, removing, 252–255
industries, drone piloting for, 9
industry-specific drones, 24
Inertial Measurement Unit (IMU), 70
influence, flying under the, 76
inspections
  drone piloting for, 299–300
  propellers, 66–68
Instagram photos, Part 107 certification and, 86
instrument flight rules (IFR), 144, 168
insurance, 25–26, 311–312
interference, lost links and, 239
International Organization for Standardization (ISO), 266
internet presence, establishing an, 307–308
interpreting sectional charts, 191–205
invulnerability, 254
iris, 129
irrigation, drone piloting for, 296
ISO (International Organization for Standardization), 266
isogonic lines, 183

J

Jamestown Regional Airport (JMS), 154
jobs, for pilots, 293–312
joysticks, 15–17
JPEG file type, 266–267

K

Kespry 2S drone, 302
Krasniak, Michelle (author)
  Social Media Marketing All-in-One For Dummies, 308

L

LAANC (Low Altitude Authorization and Notification Capability) authorization
  for night flights, 137
  obtaining, 120
  obtaining for night flights, 124

landing
  for first time, 40–42
  planes, 214–218
  runways, direction, and, 216–218
  when in doubt, land, 238
Landscape MasterShots, 62
language, Part 107 exam and, 87
latitude, 178–182
launch pad, considerations for, 39
law enforcement, drone piloting for, 9, 297–298
left-hand traffic pattern, 214–215
legs, of landing, 215–216
lens, 130
LiDAR (Light Detection and Ranging), 295, 296
lifelong learners, as a characteristic of pilots, 11
lift, 13, 244
Light Detection and Ranging (LiDAR), 295, 296
lights
  on airplanes, 135–136
  built-in, 326
  visual line of sight (VLOS) and, 105
Limited Liability Corporation (LLC), 306
lithium-ion polymer (LiPo) batteries
  about, 43
  taking care of, 71–73
livestock tracking, drone piloting for, 296
LLC (Limited Liability Corporation), 306
load factor, 245–247
loading
  about, 244
  center of gravity, 247–248
  critical angle of attack for fixed wing
    drones, 248–249
  force equilibrium, 244–245
  load factor, 245–247
  on Part 107 exam, 90–91
locations
  choosing for first flight, 38–39
  questions about, 197–199
  reading on sectional charts, 179–182
logging
  flights, 118–119
  maintenance, 118–119

logistics, drone piloting for, 300–301
longitude, 178–182
Lookup Tables (LUTs), 291
losing visual line of sight (VLOS), 330
lossy, 267
lost drones, 241
lost-link emergencies
  about, 234, 238
  handling, 239–242
  reasons for, 238–239
Low Altitude Authorization and Notification
    Capability (LAANC) authorization
  for night flights, 137
  obtaining, 120
  obtaining for night flights, 124
low angle, 271–272
Lume Cube, 106, 125
LUTs (Lookup Tables), 291

M

machoism, 254–255
macula, 131
maintaining visual line of sight (VLOS), 75
maintenance and safety
  about, 65, 243
  calibrating drone, 70–71
  changing propellers, 66, 68
  cleaning drone body and propellers, 68–69
  cleaning drone camera and gimbal, 69–70
  foundational safety practices, 74–76
  inspecting propellers, 66–68
  loading, 244–249
  logging, 118–119
  making sound judgment calls, 249–250
  models, 250–252
  removing incorrect thinking, 252–255
  storing drones, 73
  taking care of LiPo batteries, 71–73
  troubleshooting, 73–74
  updates, 66
maneuvers
  basic, 47–54

built-in, 58–59

complex, 54–58

Manual Focusing (MF) mode, 262

Manual mode, 261

mapping, drone piloting for, 294–296

Maritime air masses, 222

marketing, drone piloting for, 9

markings, runway, 214

MasterShots (DJI), using, 61–62

maximum elevation figures (MEFs), 184–185

maximum height, 100–103

Mean Sea Level (MSL), 100–102, 184

means of compliance (MOC), 114

Medication, in IMSAFE model, 251

MEFs (maximum elevation figures), 184–185

megapixels, 267

memorization, for Part 107 exam, 95

memory cards, 268

Mena Intermountain Municipal Airport (MEZ), 208

Menu, as onscreen feature, 33

Meteorological Aerodrome Report (METAR), 90, 229–230, 232

methodical nature, as a characteristic of pilots, 11

MEZ (Mena Intermountain Municipal Airport), 208

MF (Manual Focusing) mode, 262

Microdrones MD4-3000, 302

Military Operations Area (MOA), 160–161

Military Training Route (MTR), 167–170, 324–325

Minot International Airport (MOT), 152–153

misshapeness, in propellers, 67

MOA (Military Operations Area), 160–161

MOC (means of compliance), 114

models

DECIDE, 250–251

IMSAFE, 251–252

PAVE, 252

money, emergencies and, 234

MOT (Minot International Airport), 152–153

motion blur, during night flights, 127

moving vehicles

flying from, 117

flying over at night, 124

MSL (Mean Sea Level), 100–102, 184

MTR (Military Training Route), 167–170, 324–325

multispectral imaging, drone piloting for, 296

music, adding to videos, 292

myths, drone piloting, 321–326

N

National Airspace System (NAS)

about, 29, 75, 143–144

Class A airspace, 144–145

Class B airspace, 145–148

Class C airspace, 149–151

Class D airspace, 151–153

Class E airspace, 153–157

Class G airspace, 157

special-use airspace, 157–161

National parks, 121

natural elevation maximums, 184

Neutral Density (ND) filters, 273

newspapers, advertising in, 309

nicks, in propellers, 67

night flights

about, 123

airplane lights and, 136

built-in lights, 326

developing best practices for, 136–139

drone limitations for, 126–128

eyesight limitations for, 128–132

FAA regulations for, 124–125

visual problems and illusions, 133–136

nimbostratus clouds, 223

nonprofit photos, Part 107 certification and, 85

nontowered airports, 210–211

NOTAM (Notice to Air Missions), 165–166, 324–325

notations, 165

notes, for Part 107 exam, 316–317

Notes and additional information, on charts, 209

Notice to Air Missions (NOTAM), 165–166, 324–325

O

Obermeir, Barbara (author)

Photoshop Elements For Dummies, 274

obstacle-avoidance behaviors
  about, 36–38
  overdependence on, 236, 331
obstructions
  about, 185–186
  lost links and, 239
  questions about, 199–202
  wind and, 226
off-balance flights, 330
off-center viewing, for night flights, 138–139
1,200 AGL, Class E airspace starting at, 157
online payments, 311
onscreen features, of remote controllers (RCs), 33–34
optic nerve, 130
orbit maneuver, 55–56, 283–284
overdependence, on obstacle avoidance, 236, 331

P

Padova, Ted (author)
  Photoshop Elements For Dummies, 274
paid, getting, 309–310
Panorama mode, 262
Part 107 certification. *See also* TRUST certification
  about, 79–80
  certified remote pilots, 82–83
  Crew Resource Management (CRM), 92–93
  drone weight, 80–81
  exam for, 86–92
  registering for exam, 96–97
  requirements for, 83–86
  studying for exam, 94–96
  test-day tips, 315–320
PAVE model, 252
paying taxes, 312, 326
people
  flying over, 113–117
  flying over at night, 124
  flying over large groups of, 116–117
Perceive, in 3-P model, 249–250
Perform, in 3-P model, 249–250.
performance, on Part 107 exam, 90–91

personal registration, 18–19
phantom motion, 133
photogrammetry, 295
photography, drone piloting for, 301
photoreceptor cells, 131–132
photos
  about, 259–260
  angles, 269–273
  basics of taking good, 260–266
  blurry, 328
  composition, 268–269
  cropping, 274
  editing, 274–275
  exposure, 262–266
  file types, 266–267
  improving color in, 275
  improving exposure of, 275
  Neutral Density (ND) filters, 273
  resolution, 266, 267–268
  shooting modes, 261–262
Photoshop Elements For Dummies (Obermeier and Padova), 274
Photo/Video selection options, as onscreen feature, 33
Pilot in Command (PIC), in PAVE model, 252
Pilot Institute, 94
pilots
  certified remote, 30
  characteristics of, 11
  jobs for, 293–312
  recreational, 28–29
  remote, 10–11
pitch, 13–14, 15–17
Pix4D, 295
plus sign maneuver, 52–53
PNG file type, 267
POI (Point of Interest) option, 60
Point of Interest (POI) option, 60
Portrait MasterShots, 61
post-flight checks
  about, 42–43
  importance of, 63
  for night flights, 137

power lines, 39

practice and compare, 273

practicing
  basic drone flight maneuvers, 47–54
  foundational emergency responses, 237–238
  for Part 107 exam, 95

precipitation fog, 228

pre-flight checks/procedures
  about, 39–40
  importance of, 63
  for night flights, 137

pre-flights checks principle, 47

PremiumBeat, 292

pre-planned flight termination points (FTP), for
    list-link emergencies, 241

principles, flight, 46–47

Pro mode, 261

Process, in 3-P model, 249–250

profession, drone piloting as a, 8–10

professional drones, 22–23

Prohibited airspace, 159, 160

propellers
  changing, 66, 68
  cleaning, 68–69
  inspecting, 66–68
  in pre-flight checks/procedures, 40
  skin laceration from, 116

property, flying over, 112–113

prosumer drones, 21–22

Proximity MasterShots, 61

public safety, drone piloting for, 9

pull back movement, 285–286

pupil, 129

Q

quadcopters, 12, 16–17

questions/answers
  about airports, 192–197
  about airspace, 192–197
  about elevations, 197–199
  about locations, 197–199
  about obstructions, 199–202
  about sectional charts, 186–189
  about special-use areas, 202–204
  about towers, 199–202
  for Part 107 exam, 317–319

quick shots, 58

R

Radclo drone, 19, 20

radiation fog, 228

radio communications
  about, 163–164, 172
  Automatic Terminal Information Service
      (ATIS), 174–175
  Common Traffic Advisory Frequency
      (CTAF), 173–174
  control tower (CT), 174–175
  phraseology for, 175
  Universal Communications (UNICOM), 173–174

RAW file type, 267

RCs (remote controllers)
  about, 32–34
  connectivity issues with, 329
  onscreen features, 33–34
  in post-flight check, 42
  restarting during lost-link emergencies, 241
  using, 15–17

reading
  chart supplements, 208–209
  locations on sectional charts, 179–182

real estate
  ascend/descend movements for, 282, 283
  drone piloting for, 9, 294–296

recreational flyers, TRUST certificates and, 81–82

recreational pilots, 28–29

Recreational UAS Safety Test (TRUST), 29, 82.
    See also TRUST certification

rectangle maneuver, 50–51

red lights, on airplanes, 135–136

reflective tape/stickers, visual line of sight
    (VLOS) and, 105

registered agent, 306

registering
  drones, 28–30, 321–322
  for Part 107 exam, 96–97

regulations
    from FAA, 74–75
    flight, 88–89
    for night flights, 124–125
relative humidity, 228
remark, on METAR, 230
Remember icon, 3
remote controllers (RCs)
    about, 32–34
    connectivity issues with, 329
    onscreen features, 33–34
    in post-flight check, 42
    restarting during lost-link emergencies, 241
    using, 15–17
Remote Identification (Remote ID), complying
    with, 106–109
remote pilots, 10–11
replacement plans, 25–26
reporting accidents to FAA, 117–118, 324
resignation, 255
resolution, for photos, 267–268
responsibility
    about, 243
    loading, 244–249
    making sound judgment calls, 249–250
    models, 250–252
    removing incorrect thinking, 252–255
Restricted airspace, 158–159
retina, 130
Return to Home (RTH)
    about, 35, 74
    as a crash safety feature, 36
    during lost-link emergencies, 239–240
    need for, 325–326
    troubleshooting and, 42
reveals, ascend/descend movements for, 282
reversible perception illusion, 134–136
reviews
    about, 24
    for drones, 23
right of way
    about, 218
    yielding, 237

rods, in eyes, 130–132
roll, 14, 16–17
RTH (Return to Home)
    about, 35, 74
    as a crash safety feature, 36
    during lost-link emergencies, 239–240
    need for, 325–326
    troubleshooting and, 42
Ruko F11 series, 20
runway number, as a runway marking, 214
runways
    about, 211–214
    direction, landing, and, 216–218

S
safety first principle, 46
San Francisco International Airport (SFO), 146, 147
scanning, for night flights, 138–139
scotopic vision, 131
scratches, in propellers, 67
SD card, in pre-flight checks/procedures, 40
search and recue, drone piloting for, 298–299
sectional charts
    about, 101–102, 144, 158–159, 164,
        177–178, 191–192
    elevation, 183–186, 197–199
    GPS digital coordinates, 182
    interpreting, 191–205
    isogonic lines, 183
    latitude, 178–182
    locations, 197–199
    longitude, 178–182
    maximum elevation figures (MEFs), 184–185
    natural elevation maximums, 184
    obstructions, 185–186, 199–202
    on Part 107 exam, 95
    on Part 107 test, 316
    questions about, 205
    questions about airspace and airports, 192–197
    questions/answers about, 186–189
    reading locations on, 179–182
    special-use areas, 202–204
    towers, 185–186, 199–202

security, drone piloting for, 9, 301–302

seeding, drone piloting for, 296

semi-circle movement, 284–285

sensor system, 36–38

sensors, limitations of with night flights, 127–128

700 AGL, Class E airspace starting at, 155–157

SFO (San Francisco International Airport), 146, 147

shipping, drone piloting for, 300–301

shooting
  with active tracking, 59–60
  modes for, 261–262

shopping, for drones, 18–24

shutter speed, 263–264

Signal, as onscreen feature, 33

situational awareness, Crew Resource Management (CRM) and, 93

size-distance illusion, 134

skills
  development of, for careers, 304
  growing, 63

skin laceration, 116

sky conditions, on METAR, 230

Sky Life Photography, LLC, 307

slight downward angle, 272–273

small business photos, Part 107 certification and, 84–85

social media
  advertising on, 309
  setting up a presence on, 308

Social Media Marketing All-in-One For Dummies (Krasniak), 308

software
  as an expense, 311
  built-in, 289
  glitches and errors with, 327–328
  lost links and problems with, 239
  updates to, 66

sole proprietorship, 305

Soundstripe, 292

source regions, 222

special-use airspace
  Alert area, 159, 160
  Military Operations Area (MOA), 160–161

Prohibited airspace, 159, 160
  questions about, 202–204
  Restricted airspace, 158–159
  Warning areas, 158

speed
  flight regulations and, 103–104
  for Part 107 test, 317

spiral maneuver, 54–55

splitting video clips, 290

sponsorship, advertising with, 309

Spotlight option, 60

spraying, drone piloting for, 296

stable air, 224

stairs maneuver, 53–54

stalls, 248

standard day, 224

Standard Remote ID, 106–107

standing TFRs, 167

starburst effects, during night flights, 126, 127

State Farm, 26

station identifier, on TAF, 231

steam fog, 228

Stop, troubleshooting and, 42

storage
  of drones, 73
  in post-flight check, 43

straight forward, straight back maneuver, 49–50

straight up/down maneuver, 48–49

stratocumulus clouds, 223

stratus clouds, 222

Stress, in IMSAFE model, 251

studying, for Part 107 exam, 94–96

sun, visual line of sight (VLOS) and, 105

surface, 145

sustained flight, 115

sweep movement, 287

T

TAF (Terminal Aerodrome Forecast), 230–231, 232

Take off, land, return to home (RTH), as onscreen feature, 34

taking off, for first time, 40–42

task management, Crew Resource Management (CRM) and, 93

taxes, paying, 312, 326

technical proficiency, as a characteristic of pilots, 11

temperature
  on METAR, 230
  tips for, 232

Temporary Flight Restriction (TFR), 166–167, 168, 169

Terminal Aerodrome Forecast (TAF), 230–231, 232

test-day tips, 315–320

testing supplement, 164, 191, 246

TFR (Temporary Flight Restriction), 166–167, 168, 169

thinking, incorrect, 252–255

third-party plans, 25

3D models, 295, 296

3-P model, 249–250

threshold, as a runway marking, 214

thrust, 12, 244

thunderstorms, tips for, 232

tilted flights, 330

time, on METAR, 229

Timelapse mode, 262

timelines, editing video clips on, 289–290

Tip icon, 3

torque, 13

towered airports, 210–211

towers
  about, 185–186
  ascend/descend movements for, 282
  questions about, 199–202

toy drones, 19–20

track movement, 287–288

traffic, airport, 211–218

Traffic pattern, on charts, 209

transitions, applying for videos, 290

transmission range, as a reason for lost links, 238

triangle maneuver, 51–52

trimming video clips, 289

Tropical air masses, 222

troubleshooting

common problems, 73–74

flying, 42

TRUST (Recreational UAS Safety Test), 29, 82. See also TRUST certification

TRUST certification. See also Part 107 certification
  about, 18–19, 29, 79–80
  drone weight, 80–81
  in pre-flight checks/procedures, 40
  recreational flyers and, 81–82

TV, advertising on, 309

twilight, flight regulations during, 105–106

U

UAG (Unmanned Aircraft General - Small), 86

UAS (Unmanned Aircraft Systems), 28, 112, 123

ultra-high frequency (UHF) radio, 153

uncontrolled airspace, 143

Universal Communications (UNICOM), 173–174

Unmanned Aircraft General - Small (UAG), 86

Unmanned Aircraft Systems (UAS), 28, 112, 123

unstable air, 224

unusable area, as a runway marking, 214

updates
  to book, 3
  checking for, 30–31
  to firmware, 66
  to software, 66

upslope fog, 228

upwind leg, 215–216

V

valley fog, 228

VFR (Visual Flight Record), 172

VFR (visual flight rules), 168

Victor Airway, 170–171

Victor Route (VR) notations, 88

video
  about, 277–278
  adding music, voiceovers, and effects, 292
  applying transitions, 290

arranging clips, 290

ascend movement, 282–283

audience considerations, 278–279

bird's-eye rotation, 285

color in, 291

descend movement, 282–283

details for, 279

editing, 288–292

editing clips on timelines, 289–290

exporting, 292

exposure in, 291

file formats, 280

fly in/pull back movement, 285–286

frames per second (FPS), 281–282

importing clips, 289

Lookup Tables (LUTs), 291

modifying, 279

orbit movement, 283–284

planning, 278

resolution, 281

semi-circle movement, 284–285

shaky, 328

shooting more than you need, 279–280

sweep movement, 287

track movement, 287–288

visibility

  on METAR, 230

  on TAF, 231

Visual Flight Record (VFR), 172

visual flight rules (VFR), 168

visual line of sight (VLOS)

  about, 28, 323–324

  flight regulations and, 104–105

  losing, 330

  lost links and, 238

  maintaining, 75

  maintaining during lost-link
    emergencies, 240–241

  in pre-flight checks/procedures, 40

visual observer, 104, 105

visual problems/illusions, for night
  flights, 133–136

VLOS (visual line of sight)

  about, 28, 323–324

  flight regulations and, 104–105

  losing, 330

  lost links and, 238

  maintaining, 75

  maintaining during lost-link
    emergencies, 240–241

  in pre-flight checks/procedures, 40

voiceovers, adding to videos, 292

VR (Victor Route) notations, 88

W

Warning areas, 158

Warning icon, 3

warranties, 25–26

weather

  about, 221

  air masses, 222–224

  Aviation Routine Weather Report
    (METAR), 229–230

  cloud types, 222–223

  convection currents, 225

  density altitude, 224

  downdrafts, 227

  drone piloting for monitoring, 297

  emergencies with, 234–235

  flight regulations and visibility, 103–104

  fog, 227–229

  humidity, 227–229

  importance of checking, 63

  on METAR, 230

  on Part 107 exam, 89–90

  in pre-flight checks/procedures, 39–40

  as a reason for lost links, 238–239

  stable air, 224

  Terminal Aerodrome Forecast (TAF),
    229, 230–231

  tips for flying in bad weather, 232

  unstable air, 224

  wind, 226–227

  wind shear, 226

weather principle, 47

web pages, building, 307–308

websites

  Adobe Lightroom, 275

  Adobe Premiere Pro, 289

  Aeronautical Decision Making (ADM), 249

  Airman Knowledge Testing Supplement for Sport Pilot, Recreational Pilot, Remote Pilot, and Private Pilot, 164

  anti-collision lighting, 125

  Apple Final Cut Pro, 289

  Aviation Medical Examiners (AMEs), 121

  Aviation Weather Center, 229

  B4UFLY app, 35, 323

  chart supplements, 208

  Cheat Sheet, 3

  DaVinci Resolve, 289

  drone insurance, 312

  Drone Log Book, 118

  drone model database, 114

  Drone Pilot Ground School, 94

  Epidemic Sound, 292

  FAA, 30

  FAA Drone Zone, 29, 81, 118

  FAA-Recognized Identification Area (FRIA), 108

  Low Altitude Authorization and Notification Capability (LAANC) authorization, 120

  Lume Cube, 106, 125

  METAR codes, 230, 231

  Notice to Air Missions (NOTAM), 165

  Pilot Institute, 94

  PremiumBeat, 292

  recreational drones, 29

  recreational flight safety questions, 82

  Recreational UAS Safety Test (TRUST), 29

  Remote ID broadcast modules, 107

  sectional charts, 179

  Sky Life Photography, LLC, 307

  Soundstripe, 292

  Temporary Flight Restriction (TFR), 167

  Terminal Aerodrome Forecast (TAF), 230

  testing supplement, 191, 246

  YouTube, 260

  Zulu-to-standard-time converter, 231

weight

  about, 244

  of drones, 80–81

weight maximum, 18–19, 29, 321–322

white lights, on airplanes, 135–136

wind

  about, 226

  downdrafts, 227

  obstructions and, 226

  tips for, 232

  wind shear, 226

wind shear, 226

wind speed

  on METAR, 229

  on TAF, 231

workflow, 289

Y

yaw, 14–16

yielding right of way, 237

YouTube, 260

Z

Zipline, 301

Zulu time

  on METAR, 229

  on TAF, 231

Zulu-to-standard-time converter, 231

About the Author

Curt Simmons has been a technology author and teacher for the past 20 years. The author of dozens of popular technology titles, including the most recent, *Windows 11 For Seniors For Dummies*, Curt enjoys learning about and writing about all kinds of technology. Along with writing and teaching, Curt specializes in DSLR and drone photography. He owns and operates a local drone piloting business, Sky Life Photography.

Curt holds bachelor's and master's degrees that have nothing to do with technology. How's that for money well spent! His first entrance into the technology world was through a teaching job, and then an editing job, and then a book authoring job, and then a drone piloting job . . . you get the picture.

When Curt is not flying a drone or writing about technology, he enjoys travel, gardening, and fishing, and he endures the never-ending process of working on his Victorian home. He is married to Dawn and has two adult daughters and one granddaughter, who is the cutest baby ever!

You can get in touch with Curt at curt@skylifephotography.com.

Authors' Acknowledgments

No good book comes together without a bunch of people in the mix, and I'm grateful to have had great people along for the ride.

I'd like to thank my technical editor, BJ Holtgrewe, who made sure I didn't mix up the details of Class B and C airspaces (which is easy to do!). Thanks, BJ, for the eagle eye!

Thanks a bunch to my project manager and editor, Susan Christophersen, and my executive editor Steve Hayes. They both provided valuable insight that made this book better!

Also, this book has a lot of graphics to deal with, so thanks to the Production team for making everything look awesome!

Dedication

This book is for my wife, Dawn, who first said, "Why don't you just buy a drone" after listening to me talk about it incessantly for a month. Thanks for always encouraging me!

Publisher's Acknowledgments

Executive Editor: Steve Hayes

Project Manager and Copy Editor:
Susan Christophersen

Technical Editor: BJ Holtgrewe

Production Editor: Tamilmani Varadharaj

Cover Image: © Jordi Salas/Alamy Stock Photo

PERSONAL ENRICHMENT

Staying Sharp
9781119187790
USA $26.00
CAN $31.99
UK £19.99

Facebook
9781119179030
USA $21.99
CAN $25.99
UK £16.99

Guitar
9781119293354
USA $24.99
CAN $29.99
UK £17.99

Investing
9781119293347
USA $22.99
CAN $27.99
UK £16.99

Beekeeping
9781119310068
USA $22.99
CAN $27.99
UK £16.99

Digital Photography
9781119235606
USA $24.99
CAN $29.99
UK £17.99

Meditation
9781119251163
USA $24.99
CAN $29.99
UK £17.99

Pregnancy
9781119235491
USA $26.99
CAN $31.99
UK £19.99

Samsung Galaxy S7
9781119279952
USA $24.99
CAN $29.99
UK £17.99

iPhone
9781119283133
USA $24.99
CAN $29.99
UK £17.99

Crocheting
9781119287117
USA $24.99
CAN $29.99
UK £16.99

Nutrition
9781119130246
USA $22.99
CAN $27.99
UK £16.99

PROFESSIONAL DEVELOPMENT

Windows 10
9781119311041
USA $24.99
CAN $29.99
UK £17.99

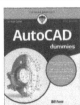

AutoCAD
9781119255796
USA $39.99
CAN $47.99
UK £27.99

Excel 2016
9781119293439
USA $26.99
CAN $31.99
UK £19.99

QuickBooks 2017
9781119281467
USA $26.99
CAN $31.99
UK £19.99

macOS Sierra
9781119280651
USA $29.99
CAN $35.99
UK £21.99

LinkedIn
9781119251132
USA $24.99
CAN $29.99
UK £17.99

Windows 10
9781119310563
USA $34.00
CAN $41.99
UK £24.99

SharePoint 2016
9781119181705
USA $29.99
CAN $35.99
UK £21.99

Fundamental Analysis
9781119263593
USA $26.99
CAN $31.99
UK £19.99

Networking
9781119257769
USA $29.99
CAN $35.99
UK £21.99

Office 2016
9781119293477
USA $26.99
CAN $31.99
UK £19.99

Office 365
9781119265313
USA $24.99
CAN $29.99
UK £17.99

Salesforce.com
9781119239314
USA $29.99
CAN $35.99
UK £21.99

Coding
9781119293323
USA $29.99
CAN $35.99
UK £21.99

dummies.com

dummies
A Wiley Brand